LLAMA
ULTIMATE

TRAINING, FEEDING, PACKING, HUNTING FISHING AND CARE

DOYLE MARKHAM

Snake River Llamas

Copyright 1990 by Doyle Markham and Snake River Llamas. All rights reserved. No part of this publication may be reproduced, stored in a retrieval system or transmitted in any form or by any electronic means, mechanical, photocopying, recording, or other, without written permission from the publisher.

ISBN 0-9628326-0-X

Library of Congress Catalog Card Number 90-92191

First Printing: November 1990

Published by: Snake River Llamas
              7626 North 5th West
              Idaho Falls, Idaho 83402

Disclaimer of Liability:
The author and Snake River Llamas are not liable for any loss or damage allegedly caused directly or indirectly by using information in this book. Use the methods, procedures and products at your own risk.

LLAMAS ARE THE ULTIMATE

TRAINING, FEEDING, PACKING, HUNTING, FISHING AND CARE

DOYLE MARKHAM

TABLE OF CONTENTS

CHAPTER	TITLE	PAGE
1	WHY A LLAMA?	1
2	LLAMA HISTORY	8
3	TRAINING - WHEN DO YOU START?	15
4	TRAINING THE WEANLING	18
5	TRAINING LLAMAS TO TRANSPORT	24
6	TEACHING YOUR LLAMA TO KUSH	29
7	BASIC TRAINING FOR THE ADULT	32
8	TRAINING - PACKING WITH LLAMAS	38
9	AT WHAT AGE CAN YOU START PACKING YOUR LLAMA?	50
10	HOW MUCH WILL THEY CARRY?	53
11	ON THE TRAIL	59
12	PACKING WITH A PACK STRING	70
13	TRAIL AND CAMP ETIQUETTE	79
14	PICKING A PACK LLAMA	84
15	TRANSPORTING LLAMAS	94
16	COMMERCIAL PACKING	102

17	SADDLES, PANNIERS AND THEIR CONTENTS	110
18	THE LOST LLAMA	125
19	CATCHING LLAMAS	133
20	FEEDING LLAMAS AT HOME AND ON THE TRAIL	138
21	BARNS, SHEDS AND WINDBREAKS	150
22	FENCES	156
23	PASTURE MANAGEMENT	161
24	WITH LLAMAS, PACKING GAME IS FUN	171
25	HUNTING AND PACKING WITH LLAMAS	185
26	BACK COUNTRY FISHING WITH LLAMAS	197
27	LLAMAS, THE ULTIMATE SHEEP GUARD	206
28	FLYING AND JET-BOATING WITH LLAMAS	218
29	CHOOSING A LLAMA VETERINARIAN	229
30	HEALTH AND CARE	238
31	EMERGENCY AID ON THE TRAIL	248
32	CATS AND LLAMAS	255
33	SELECTED PUBLICATIONS AND INFORMATION ON LLAMAS	259
	AT THE TRAIL HEAD	268
	ACKNOWLEDGEMENTS	269
	INDEX	270
	MEET THE AUTHOR	286

DEDICATION

I know of no one who loves llamas more than my wife, Cherrie Markham. Many times I have seen this love and humility displayed by moist eyes or a wet cheek over seeing a new cria born in our pasture. I've heard her adoration as she talks to llamas both at home and on the trail. They may not understand someone else, but I think they understand what she says. Many times on the trail, I've witnessed this love and seen the pride on Cherrie's face as she praises a llama for a difficult task or for carrying a big load on a long day. I have seen her commitment to llamas; after grading math papers for her 150 students, as Secretary/Treasurer of the Northern Rockies Chapter/ILA and co-editor of the newsletter, she has worked on a llama newsletter late into the night. As a partner in Snake River Llamas, I've witnessed her kind concern for the care and training of our llamas and her continual hard labor to care for and improve our llama farm. Although we have lived in town in a comfortable house for 19 years, her love for llamas exceeds her emotions to live in a home where pleasant memories abound, especially those of two wonderful children growing up. So my next project is "....come, come away with me and I'll build you a home in the meadow" - in this case, right in the middle of our best pasture where she can see her beloved llamas out of any window! This book is dedicated to Cherrie.

CHAPTER 1

WHY A LLAMA?

If you own llamas, you surely have been asked this question or one similar to it, such as "What are they good for?" However, if you own llamas, you likely can't think of a life without them. If you don't own llamas, this chapter may provide a little insight into people you know who are a little crazy about llamas.

One of the most important reasons for owning a llama is that they enrich and help you enjoy life. Llamas are just plain fun to own! Most people think they are beautiful, with their luxurious fur, large eyes and oversized eyelashes. Their looks, and particularly their tranquil attitude and behavior, make them interesting and pleasing to see. Their graceful movement is relaxing to watch and each one acts as if they are royalty. They display themselves as proud beings. Llamas do wonders for the person who is involved in the hustle and bustle of life. A few minutes with the llamas after work has a calming, soothing effect on the nerves.

Also llamas are very useful. I love to pack into the high country of our western forests and wildernesses during the summer. So to me, one of the best things about owning llamas is that they are wonderful packers who will easily and contently pack about anything that we ask them to pack. I'm a hunter who hunts with llamas. Packing wild game is just another load to them, even to one who has only packed once or twice. They will pack up to 33% of their body weight and some of the better packers weigh over 400 pounds so they can pack a lot of weight. Many can carry even more than the

WHY A LLAMA? 2

Llamas enrich our lives, are fun to own and make safe animals for kids to handle.

A wonderful Christmas gift! Nearly everyone seems to enjoy llamas. Llamas help us to relax and to be happy.

WHY A LLAMA? 4

recommended amount for short distances. The entire family will enjoy packing into the wilderness, and during the summer, its a special treat that everyone will enjoy.

One reason packers like llamas is that they are so smart. Rocky and Himalaya help me choose the proper route.

They are always alert and seem to enjoy packing. They have remarkable eyesight and often spot game long before I do. We use this trait to good advantage when we are hunting. I have positioned some of them to watch the other side of a ridge where I hunt elk. I watch my side and then occasionally glance over at the llamas to see if they have spotted anything on the other side of the ridge.

Llamas are also one of the easiest animals to care for. Although you'll want to spend a lot of time with your llamas, their daily needs can easily be provided in only a few minutes. They are generally disease free, and they seldom require medical attention. Llamas are also economical to own. A

LLAMAS ARE THE ULTIMATE 5

Llamas are easy to raise and to train. The females are excellent mothers.

mature adult will feed seven to ten days on a bale of hay. Since males start around $1,000 to $1,500 and can live between 20 and 30 years, the cost per year for purchasing and maintenance is very reasonable. When hay was selling for $40 a ton in our area, a friend of mine kept records and it was cheaper to feed three llamas than it was to feed his Brittany dog. Since their food requirements are low, a small suburban pasture can be used to provide grazing for two or three llamas.

WHY A LLAMA? 6

Llamas are easily trained and most people can do a good job of training their own llama. If you're interested in packing with llamas, 4,500 - 6,000 years of experience packing in South America makes them a natural when it comes to packing. Llamas usually pack a load very well even on the first time in the mountains. We often begin packing in the mountains after having a saddle and panniers on them only once. Llamas are the easiest packing livestock to train and the safest to handle. Therefore, llamas make excellent 4-H projects and pets for children.

Llamas can be hauled in pickup stock racks, trailers, vans and camper tops. They quickly learn to load and generally lie down when you're driving. You can even train a llama not to mess in your trailer or van. A trained pack llama can jump into the bed of a four-wheel drive pickup with impressive agility and grace.

Llamas make good neighbors. Almost any type of fence will keep them home. A farmer friend of mine indicates that he has owned all types of livestock, and the llama is the easiest on fences of all of them. Llamas are also quiet; even with a large herd, you will seldom hear any sounds. They also do not have an odor. Their pellet feces do not smell and can even be used as an indoor fertilizer for house plants.

Llamas are also good investments. The economics of llamas has probably attracted the most people to llamas. Currently most people are making over 20% annually on their investments. Demand generally exceeds supplies, thus an investment in female llamas pays big returns.

Another reason for owning llamas is their wool. Llama wool can be obtained by shearing or by brushing the wool. Clothes made out of llama wool are luxurious and warm. Sheep and goat producers purchase llamas because they are extremely effective guards against predators. Many sheep breeders lost over a hundred lambs per year prior to purchasing a llama to guard their sheep. After purchasing a llama, their predator losses dropped to zero.

I can't imagine a life without llamas. In the winter, it's a special treat to drive into our barn area to feed llamas where we keep our geldings and young studs and see the males charge round and round the truck. I also can't imagine not being greeted by Himalaya each time he sees me. He comes to us, smells our face and walks along beside us every time we go into the llama area.

A word of caution is necessary about llamas! People just don't own llamas; they usually become very attracted to and involved with their llamas. This attraction isn't something that is easily explained, but it quickly occurs. I'm sure you'll find that llamas will greatly enhance your enjoyment. This book is co-dedicated to the growing number of people who are owned by llamas. If you're not one of these lucky folks, we hope that you soon will be.

CHAPTER 2

LLAMA HISTORY

Although we often think of llamas as a fairly recent import into North America from South America, llamas really are North American natives since llamas inhabited North America for many millions of years before they inhabited South America. Current archeological evidence traces the beginning of lama (members of the genius **Lama**) history in North America at approximately 35-million years before present. The family camelidae, which includes both camels and lama species, had much of their evolutionary history in North America. Approximately three-million years ago, camels migrated north across the Bearing Sea land bridge that existed when the seas had less water because of frozen ice and snow on land. The bridge provided a path into Africa and Asia, which contain modern camels as we know them today. Lamas, on the other hand, went south into South America during this same time period. Although these camelids were probably not the same as the animals we recognize today, they did provide the ancestors of our present camels, guanacos and vicunas.

Approximately 10,000 to 12,000 years ago, dozens of species of large North America mammals became extinct. Among these were saber tooth tigers, woolly mammoths, horses, sloths, camels and lamas. What caused these animals to become extinct? There are several theories. At least in Idaho, man made his appearance just before this time. Recent archeological evidence clearly demonstrates that early man dined on camelids, particularly camels. One prominent theory is that man caused these animals to become extinct or at least hasten the process by hunting these species to extinction.

LLAMAS ARE THE ULTIMATE 9

The extinct camel, Camelops hesternus, and the extinct llama, Hemiauchenia macrocephala, occurred in North America. Hemiauchenia fossils have also been found in South America.

However, it is hard to conceive that a small population of early man could have hunted and killed so many and so large of animals in a geologically short time.

Several other theories exist such as the climate changed significantly so that the animals could not tolerate the change or more likely the climate caused a significant change in vegetation which the animals could not tolerate. Some wonder why they just couldn't have migrated to a better suited climate. Another theory is that some catastrophic event such as a comet coming close or an exploding star caused the animals to die, perhaps by destroying part of the ozone layer that protected them from the sun's ultraviolet rays. Destruction of the ozone layer is a concern that has also been expressed in modern times because of atmospheric pollution.

LLAMA HISTORY 10

The vicuna is the smallest of the *Lama* species. This species also has the least number of individuals.

LLAMAS ARE THE ULTIMATE 11

These theories may or may not be true, but for whatever the reason, lama species became extinct in North America. However, lamas flourished in South America. There are presently four recognized species of the genus **Lama** in South America. They are the guanaco, **Lama huanacus**, the vicuna, **Lama vicugna**, the llama, **Lama glama** and the alpaca, **Lama pecos**.

The vicuna is a small camelid that usually lives above 17,000-feet elevation in the Andes Mountains. The wool of the vicuna has fine fibers, and clothing made out of vicuna was reserved for royalty when the Incas controlled much of South America. Disobeying this law was punishable by death.

In 1532 the spaniards invaded South America. Reports indicate that up to 80,000 vicunas per year were destroyed for their hides and wool. In addition, the spaniards introduced their domestic stock, primarily sheep and alpacas, into the vicuna's habitat. In the 1960's reduction in their numbers continued as poachers slaughtered this magnificent resource, and this species was on the verge of extinction. The establishment of vicuna reserves and increased protection saved the vicuna. Currently between 100,000 and 200,000 animals exist, primarily in Peru but also in Chile, Bolivia and Argentina. Now they are once again periodically captured, sheared and released.

The guanaco usually is found between sea level and 12,000-feet elevation. Compared to the vicuna, it covers a wide area. Its normal range is dry, with hot days, cool nights in the summer and cold in the winter. A vast portion of their rangeland is similar to what we classify as steppe or desert in North America. The guanaco was one of the most important animals to the South America indians as they provided hides for clothing and shelter and meat to eat. Probably between

The guanaco provided food, clothes and shelter for South American natives.

500,000 and one million guanacos are in South America at the present time. Both the vicuna and guanaco are cinnamon-brown.

The two lama species, llamas and alpacas, were probably not ever wild animals. They were likely created by South American natives some 4-6,000 years ago by selective breeding of guanacos and vicunas. The alpaca is a small, woolly creature whose primary purpose in South America is for wool production and secondarily for meat. The alpaca is the most abundant lama species in South America, mostly in Peru and numbering in the millions. Although alpacas come in a variety of colors, white is the preferred color, and alpacas with other colors are generally harvested for meat in South

America. White, as it is for sheep in North America, is preferred since white wool can be dyed any color. Alpaca wool exports are an important source of revenue in South America.

Alpacas in South America provide wool and meat.

The llama was primarily used for packing in South America, with meat and wool secondary uses. In the 15th century, the Incas of Peru controlled 2,500 miles of South America. Llamas carried military and domestic goods through the empire. Mining operations also used thousands of llamas. One herder controlled eight to ten packing llamas by herding them. The Incas used many thousands of packing llamas. Llamas were found primarily above 8,000-feet elevation in Peru, Argentina, Bolivia and Chile. In these harsh climates, temperatures differ greatly between night and day.

LLAMA HISTORY 14

Llamas were also important social and religious symbols. White llamas usually preceded royalty. Also white llamas were usually the ones sacrificed in their religious ceremonies. Often the lungs were inflated and the future was foretold by reading the pattern of blood vessels on the lungs.

Llamas still serve remote areas as packers. However, with the increase in vehicles to transport freight, currently llamas are used primarily for meat production. The llamas of South America are smaller than their North American counterparts. At least one South America country is providing studs for leasing to llama breeders in an attempt to increase the size of their llamas and to help prevent inbreeding, which is a major problem in portions of South America.

Although we aren't sure when llamas were first imported back into North America, Eric Hoffman (Llama Life 1988:5) has researched some of the earlier records. He located 30 zoos and 18 private parties which had llamas prior to the 1930 ban on the import of llamas because of hoof and mouth disease. There likely were many others. The earliest birth was reported for 1870, and one eastern zoo reported 41 births between 1878 and 1928. Although we can't say how many llamas or how many zoos and farms had llamas during this early period, there were many of them and likely a good gene pool existed. Later, in about 1928 Randolph Hearst began to collect and breed llamas and created one of the better-known early llama herds. It is thought that Hearst originally purchased his stock from Germany and South America.

Today llamas and alpacas are located throughout the United States and Canada. They have adapted well to their ancestral home.

CHAPTER 3

TRAINING - WHEN DO YOU START?

We have a rule at our farm concerning our crias (baby llamas): they are to be admired but not touched until they are at least five-months old. There are few exceptions to this rule. We do handle them on the day they are born to obtain weights and to provide other first-day care. We also handle them briefly when we give them shots or need to move them to another pasture. Other than these few times, we don't touch our babies, especially the males.

Over-handled baby male llamas can cause significant behavior problems in later life. The most severe cases occur when male babies are bottle fed. This may result in the so called "berserk syndrome." Sue Rolfing has described the cause of the berserk male syndrome as ". . . excessive human contact that interferes in a baby llama's normal socialization process to the point that it cannot properly distinguish between llamas and people. It usually is first manifested as the animal approaches sexual maturity, and becomes aggressive and even violent towards humans." Excessive handling of females resulting from bottle feeding also causes them to be pushy and obnoxious. In the most severe male cases, involving bottle raised males, sometimes the only solution is to put the animal down. Gelding bottle raised males is absolutely necessary.

I am certainly not implying that periodic handling of a baby male which is not bottle fed will result in berserk male syndrome. However, over handling of any male or female baby can result in an animal that has less than desirable behavior when it is grown. The hands-off approach of babies

TRAINING - WHEN DO YOU START? 16

This baby or cria has several months to wait before beginning his training. Training for young llamas usually begins at five or six months of age.

LLAMAS ARE THE ULTIMATE 17

may be too conservative for many; however, it does result in well-adjusted, friendly and cooperative llamas when they are grown and without a hint of being pushy, obnoxious or dangerous.

When females are over-handled when they are young, they are often the ones that tend to spit or become aggressive when you pick up their babies and become aggressive or spit when they are haltered.

Although we do not handle our babies, an important part of our training program does take place before the babies are five-months old! We make sure that we are near the babies nearly everyday. Even if they are in a pasture, we walk near them. Also when we are feeding hay, we make sure that the feeder is in a position that forces us to carry hay into the pens and pasture rather than pitching the hay over the fence into a feeder. Thus we are in close contact with them. In this way, the babies become familiar with us and eventually lose most of their fear. Often they will approach us to within a couple of feet. We think that this is an important training technique that makes future training easier.

Some training programs do recommend some minimal contact with babies before they are five-months old. Some suggest touching the babies with a leather wand or dowel. However, most llama trainers suggest that you either do not touch them or be in contact with them only a few moments per day.

CHAPTER 4

TRAINING THE WEANLING

In the previous chapter, the subject of when to start training your llama was covered, and a conservative approach of not touching male babies until they were weaned was discussed. We wean our babies at five months of age. Some breeders wean a month or so later, and a few wean a little earlier. This is a good time to start training your little llama. Their smaller size at this age makes them easy to train.

A well-trained weanling makes a great impression on your buyers when they come to pick up a llama. They think this youngster "obviously has a good personality" or a "calm and tranquil personality" which first time buyers, in particular, like to see. You'll feel better about selling a trained llama. It is just as important to train one if you plan to add him or her to your own herd.

Let's assume that we have just weaned our babies (crias) or you have just purchased a weanling. Many breeders do not train any of their weanling llamas. As a new llama owner, don't be surprised if you have to do all the training. We have been around them often but have seldom touched them except for shots and moving them. Now for the hardest part of training: catching the little rascals. If you have planned your facilities well, you'll have a small catch pen in one corner of your corral or pasture. A catch pen is highly recommended for catching all ages of llamas, and it makes the training of babies a much easier job. Often training of youngsters works best if two people are involved. Try to limit your training to 5- to 15-minute sessions. We usually work with our youngsters only once or twice per week because of other commitments. You can set your own schedule.

If the catch pen, barn or stall used to confine the youngsters is small, some people prefer to catch a particular llama several times, stroke the neck, talk softly to the llama and then let the llama loose. After repeating this several times in two or three sessions, the llama soon learns that being caught isn't all that bad. Then add a step by putting on and taking off the halter several times. This is a gentle and easy method, and one that you may wish to use. We, unfortunately, just don't have the time, and other methods produce equally manageable llamas.

If you do not have a small catch pen, a corral will work. Catching young llamas in a pasture is not recommended as it can be time consuming and frustrating. Youngsters are easily caught in the corner of a corral. With a short rope between two people (or if one person is doing the training, one end can be tied to one side of a corner), a young llama can be worked into a corner of a corral. The llama commonly faces into the corner to get as far away from you as possible. If two people are involved, one person stands behind the llama so that he cannot run away while the other one stands by the front shoulder and restrains the front part of the llama and puts on the halter and lead. One person can accomplish the task of restraining a llama by pushing the animal against the side of the corral.

We use either a nylon halter with the fastex buckle or a small leather halter with our young llamas. If you use the nylon halter with the fastex buckle on only one llama, it can be put on and taken off quickly. However, if you are working with several different sized llamas, adjusting the halter between llamas is time consuming, and frankly a pain in the rear. The leather halter can be quickly placed on several llamas without

TRAINING THE WEANLING 20

any adjustment. We have not had good luck with nylon halters with the metal snap clips. The ones we have ordered have had flimsily clips which easily broke. It is important to make sure that your halter is the right size and is adjusted correctly so that you do not restrict breathing.

Our approach to training youngsters is more direct. Once we have captured an animal for the first time after weaning, we put on a halter and a lead and release him or her from physical restraint. When a llama discovers that he no longer has his freedom, he sometimes panics for a few moments. He may buck and jump so it is best to have plenty of room without obstacles so that he cannot get hurt. This usually lasts only for a few minutes. However, as pressure is increased on the lead, the llama sets his feet as a sure sign that he has no intention now or in the future of following you anywhere. This is where it is nice to have another person working with you. One person pulls on the llama with the lead while the other person follows along behind the llama to encourage him to go forward. We may only cover 50 to 100 yards during this first encounter.

Individual llamas differ as to how well they adapt to being led or pulled around. Usually at the beginning of the first session, the typical llama will literally have to be dragged and pushed everywhere. Usually by the end of the first session, the llama may alternatively walk a few paces then defiantly set his feet to stop all forward progress. When the llama stops and refuses to go forward, it sometimes works best to pull the little guy or gal at a angle. This tends to put them off balance, and they step forward. I think that alternate tugs or pulls sometimes works better than a continual pull.

Stop occasionally during the training and let your llama rest. I think it helps to talk to the llama during the entire training

Young weanling llamas usually learn to lead within a few sessions.

session. Following other llamas around in the corral may assist in getting the llama to move at a steady pace.

To end the first session and each subsequent session, we approach the llama while shortening the lead until we can restrain the animal by the neck and shoulders as we stand beside him facing the same direction as the llama. We then gently stroke the animal's neck, back and legs, quickly moving our hand. We often pat rather than stroke and quickly move the hand to different parts of the body. Usually, llamas are less sensitive to the pat than the lingering touch or stroke. We even pick up the legs on this first time encounter as the animal is still thinking about his experience with the lead and halter and usually is not all that concerned about the legs. We pick up each leg and gently let it down. We always run our hands under the llama to simulate saddle straps. On the females, we touch the mammary gland area which is one of the more sensitive places on a llama. The reason for doing

TRAINING THE WEANLING   22

this is obvious if you have ever had to touch the teats of a 400-pound female and thought about all that muscle in those big thighs being used on you. When I have to examine the utter of a big female, I sometimes wonder if Cherrie shouldn't be doing it. We also touch the ears, face and lips before releasing them. Then we gently and slowly release the halter and let the animal loose. Remember the entire session lasts only for 5 to 15 minutes for each llama.

Desensitizing the legs and body of a newly weaned llama is usually quickly accomplished. This llama is just six-months old, but weighs 182 pounds.

Generally within three or four sessions the individual animal is following reasonably well. At the end of six or so sessions, the average young llama easily follows you and allows all four legs to be picked up. Once a llama leads well and allows touching over his body and allows his feet to be picked up, there is no need to continue this routine type training. Llamas are smart and may become bored with a lot of repetition and may react negatively to doing the same things over and over. Once an animal has learned to lead and to permit handling of the legs, repeating these lessons once every month or two is generally all that is needed to reinforce previous lessons. Surprisingly, once they have learned to lead and to be touched, they retain the information very well with refresher sessions.

CHAPTER 5

TRAINING LLAMAS TO TRANSPORT

Another easily taught lesson for young llamas is loading in a trailer or truck. If the llamas you are teaching already lead and handle easily, teaching them to load is a simple task. During a previous December, Cherrie and I taught five llamas from five to six and one-half months old to load in a trailer in one and one-half hours. These llamas were not really trained to lead all that well; however, they still learned quickly. Of course, they should have additional lessons. Basically, it proved the popular opinion that, "Once you've shown a llama how to do something three times, they know it."

Our procedure involved the two of us. The rear of our trailer is about two feet off the ground. We place a hay bale on the ground at the end of the trailer with the hay partly under the trailer to make their entry easier and to prevent their legs from going under the trailer. One of us pulls the llama into the trailer with the lead while the other one picks up the front legs and puts them on the hay or trailer and assists by pushing on the llama's rear. On the fourth attempt, every one of the llamas jumped into the trailer with only a firm pull on the lead and no other assistance. On the fifth try, the llamas jumped into the trailer with only a little tug on the lead. Young llamas that have mastered the leading better will learn even faster. It apparently doesn't leave the llamas with much fear of the trailer. We put a young female in the trailer that needed to be taken to the vet for a health certificate at the end of that particular training session. One of the young llamas that we had just taught came and jumped into the trailer on his own to see what was happening.

If you are using the typical stock trailer where the rear of the trailer is usually less than 12 inches from the ground or if you can park your trailer or truck so the rear tires are in a low place or the rear is against a mound or hill, teaching young llamas to load will be even easier.

Llamas quickly catch on to loading. Jasper, a 400-pound gelding, jumps into our truck with little effort.

We have seen yearling llamas watch us load another llama several times, then step in without hardly any effort on our part. We have also seen an eight-month old llama jump in a trailer for the first time without a halter on just to see what was happening with another loaded llama.

Adult llamas can be taught the same way; however, if they resist, the effort required may be greater because of their size and strength. You may have to pick up a foot and put it into the trailer for the first time or two they are loaded. If llamas have been worked a few times, they often learn the loading process more quickly than do the weanlings.

TRAINING LLAMAS TO TRANSPORT 26

Once a llama has jumped into a trailer or truck at a low height, the trailer or truck can be moved to more level ground. However, don't expect weanlings to be able to jump into your truck from level ground. However, studs and geldings should be able to jump from a level area into whatever you usually haul them. Even a big four-wheel drive is no problem for most males.

Once while hunting in Montana, I shot a moose in an area where I could not get the trailer within 10 miles. I needed to load the two studs and two geldings I had with me in the stock rack on the three-fourth ton four-wheel drive. One of the llamas, Churchill, had been purchased the previous summer and had only been caught one time in his life when we picked him up. We worked with him a few times, put a saddle on him once at the farm and then took him on a two-day hike. This was only his third trek with us. This relatively inexperienced llama had been in a llama trailer only a few times and had never been in the back of the truck. I couldn't back into a hill to make loading easier because of the deep snow. Therefore, they had to be loaded on level ground. I tied Churchill close to the truck, and he watched the other males jump into the pickup. When it was his turn, Churchill jumped up into the truck without the slightest hesitation. Llamas can learn to jump into other trucks and trailers very easily.

There are at least three precautions in teaching llamas to jump into trailers. Often while learning, llamas jump unnecessarily high when entering a trailer. Because the rear entrance of a trailer is typically higher and wider than the side escape door, it is generally safer. Llamas sometimes hit their head or jump into the sides of the opening if the side door is used for teaching purposes. Some precaution must be taken to insure that the front legs do not extend under the trailer when pulling and pushing on the llamas to get them to enter. This usually isn't much of a problem if you start with a small jumping height into the trailer. As mentioned previously, bales of hay or a piece of plywood tied to the rear of the trailer help prevent this from happening on higher jumps. The third precaution is to make sure you have a halter and lead that are well made and will not break. All halters and leads are adequate if they are not worn - right? Wrong! We have purchased a number of halters for both adult and juvenile llamas that broke under an unreasonably low tension. A breaking halter or lead can be dangerous for both you and your llama when you are tugging on them.

Occasionally one runs into the situation where you either purchase or agree to transport a large adult female or male that is not trained to load and may have never loaded or has been loaded only once or twice in their lives. Loading these creatures can sometimes provide you with a bit of a problem. Some may even provide you with a real challenge. For these types of animals, two people certainly work best. Generally, the lower the entrance the easier it will be to load these critters so try to park somewhere near a mound of some sort. In nearly all cases, you can drag the reluctant critter over the threshold before she really knows what is happening. If two people are involved, one can push while the other one pulls on the rope or lead. We have never been kicked by pushing on a llama as it is being pulled forward, and we have bought and/or transported lots of llamas. I can't absolutely guarantee it won't happen to you, but I'd be surprised if it does. If you have been kicked in this situation, I'd like to know about it.

Some may simply refuse to even try to load and lay down when they get close to the trailer or truck. They may even decide that they went this far with you, and as far as they are concerned they don't want to go with you anywhere else, not

TRAINING LLAMAS TO TRANSPORT

now or ever, so they will just permanently lay where they are until you go away. Of course you, being of superior intelligence and perhaps even knowing more profanity than they, can certainly not pass up this challenge. In all but the most obstinate cases, this is usually only a temporary setback. There are several methods that one can humanely try that are somewhat successful. The first involves pulling straight up on the lead (perhaps even grasping the critter around the neck and lifting at the same time). On the trail when a young packer lays down, charging the animal and hollering usually gets the critter up. So you may want to try charging, shouting and throwing your arms about as you lift up on the lead. I'm not sure whether the shouting will help or not on loading, but it will certainly make the entire operation more entertaining to those on the sidelines or your helpers. It may also relieve some tension. Even if the shouting doesn't help get the llama up, it obviously has some good side effects. Rolling the animal to one side or lifting on the rear can also do the trick. Usually one of these actions or a little time or a combination of these will get the animal back on his feet.

If one person is trying to load a particularly difficult animal, a long lead can be used effectively. Somewhere near the inside rear of your trailer, there should be a support or loop of some type that you can use to slide the rope through to help hold the llama near the trailer while you descend from the trailer, still pulling on the rope. Now you can both pull and push to encourage the animal to enter.

Most llamas learn the loading lesson quickly and usually first-timers can be loaded fairly quickly. Teaching llamas to load will give you an appreciation for how easy llamas are to teach.

CHAPTER 6

TEACHING YOUR LLAMA TO KUSH

Many beginning llama folks have heard of llamas kushing or lying down on command. I have even seen breeders attempting to teach weanling llamas to kush on the first lesson. Some llamas have very sensitive legs, and they have a tendency to lie down when you touch them. Therefore, you do not want to reinforce this behavior by teaching your llama to kush before they are well trained. I don't teach mine to kush until the need arises, and the llama has mastered well the previously discussed training. I suggest you do the same.

Teaching a llama to kush is generally easy if the llama is well trained and trusts you. For example, one Christmas we put a new gelding into our church's live nativity pageant. Of course when the three wise men and their llamas approach the manger, all three llamas are supposed to lie down. Well, our new actor had not been taught to kush. About 30 minutes before the first performance, we began his first "kush" lesson, probably getting him to lie down a total of five times. On the first time before the crowds, he was a little rusty, but he did ok. By the third performance that first night, he was doing fine.

One person can teach a llama to kush, but it is easier with two people. One person pulls the lead downward causing the head to go close to the ground. Putting the lead under the bottom rail of your corral and pulling up on it is usually helpful. Squeezing a leg with your hand (usually, you may have to actually lift a foot) while pushing down on the shoulders will generally get the animal to go to his knees, then running the

Teaching a llama to kush. Usually it is easier with two people.

hand along the stomach will generally get the animal to finish lying down. While working with the llama say "kush" several times so he begins to associate the action with the command. Some people have excellent luck by just pushing firmly on the top of the shoulders while pulling the head and neck downward, thus avoiding the possibility of teaching your llama to lie down any time you mess with his front legs. I haven't experienced any tendency for them to lie down when handling the leg after teaching them to kush. However, please remember that we teach them to kush only after they are used to us handling their legs. At this point praise the llama generously. After about the third time, he'll get the idea and will usually kush when you give the command and pull down on the head.

Most of my packers do not know how to kush on command as I just haven't seen a need to teach them. It really isn't necessary and the reader is cautioned about teaching "kush" too early in your training program.

CHAPTER 7

BASIC TRAINING FOR THE ADULT

Many breeders do not adequately train or even attempt to train their llamas. Therefore, you may find that you have purchased an adult that has no training at all. For you the buyer, this may have some advantages and disadvantages. The disadvantages may be obvious when you try to catch the llama, put a halter on, lead or load him. In some cases, the llama may have had little if any contact with people so he may be afraid and may not be approachable. Perhaps you'll want to look at the bright side. At least, an untrained animal usually will not have developed bad habits from being improperly trained.

You may think that purchasing a young adult or weanling that hasn't had some type of training is a rare occurrence, and something that you will not encounter often. Surely most llamas for sale have been taught to halter, lead and load. "Well, we haven't worked with him much" often means "We've never taken the time to train him even once." The majority of the weanlings and young adults, both male and female, that we have purchased have had little or no training. So don't be surprised if your newly purchased animal has not been trained. I don't mind purchasing untrained young adult males, and we sometimes prefer that they not be trained. They are easy to train, and none of them become pushy, obnoxious or kick because of bad training or over handling as babies.

We've already discussed methods of training weanlings. In this chapter we will briefly discuss how to train an adult male.

We'll assume he is completely untrained but otherwise a normal animal. Often training an adult is easier and less time consuming than training a weanling. They catch on quickly. However, you'll need to be more cautious with an adult. An adult male has awesome strength. Combine this strength with fright, and you have a real strong animal on your hands that deserves some respect.

If your new animal has not been around people, you may wish to start by visiting him several times in a small corral before you actually start your training. It's a good idea to isolate any new purchase from the rest of your herd until you are sure that he is not carrying external parasites or some contagious diseases. During these few days, the llama may be somewhat apprehensive about his new surroundings. So this is a good time for you to get acquainted with him. Talk to him as you feed him each day. An important part of any training is the rapport and trust between the animal and the trainer. Use a soft, non-threatening voice as you talk to him.

As with weanlings, adults are easier to catch if herded into a small corral or catch pen. When we purchase an adult male that needs to be trained, we leave the halter on him when he is unloaded on our farm. We prefer to teach the haltering process at a later date after he is partially trained. If there are two of you, catching the animal will be much easier.

Although it is rare for our llamas to kick, I don't approach a strange male directly from the rear, and I don't think you should either. We often use a rope to herd an animal into the corner of the corral. In a small corral, two leads attached together will serve the same purpose. Once in the corner, one person can slip in beside the animal and grasp the halter while putting a lead on the halter.

Later, whenever possible, you should always remove the halter after every time you use or train your animal. The halter may get caught on a fence or other obstacle. In addition, removing the halter prevents areas from becoming tender or rubbing the hair off in some places.

WARNING: A llama that is frightened may panic and try to break away from you. Usually an adult llama can be restrained with the lead, providing he does not get his back toward you. If he does get his back to you and has panicked, depending upon your size and strength, be prepared to let go of the lead or you may get dragged around the corral. Occasionally this will happen when you put the saddle on a llama for the first time. Although it is rare for a llama to panic, except in these two circumstances, the same advice is valid for any large male llama that panics for any reason. They may be too strong for the smaller to average person to restrain if they panic and are facing away from you. It may be safer and easier simply to let go and catch the llama again. The beginning llama owner should not be overly concerned about this as it almost never happens once an animal is trained. I have seldom had a female (and we have purchased many adult females that were never handled in their lives) that we could not restrain with the lead at all times.

For this reason, a person working with a new adult male llama should never wrap the lead around his hand or attach it to his body in any way. It is a safe practice to always hold a lead in such a way that you can easily drop it, and this advice applies to leading any llama, trained or not.

An excited, frightened male that does not have a collar or halter on is a little more difficult to catch. However, the procedure for two people works great, usually without any difficulty. We herd the animal into a corner. One pushes the animal on the rear to keep him in the corner while the other one leans against the animal's front shoulder and places a rope or lead around his neck. Although it is usually not necessary to do, you can loop the ends of the lead around a corral pole to help you restrain the animal. I put my arms around his neck, facing the same way he does, with my body standing at his shoulders. Each hand has an end of the halter, and I bring the halter slowly up under his nose. A shorter person may have trouble reaching the nose of a tall male as the general reaction of the male is to put his neck and head as high in the air as he can. Usually by stepping on a corral pole one can reach the head. Although it is usually not necessary, you may need to tie his neck to the pole with a quick release knot or by using a quick release snap in order to climb up on the corral poles. You may need to restrain the neck and head movements in order to put the halter on. Usually this can be done with the arms and your body as you reach around the neck. It is usually easier to halter any llama from the side rather than to stand in front of him. Try to avoid having your head at the same level as his head. They are not prone to jerk their heads to the side; however, if one jerks his head to the side, it may collide with yours. It seldom happens. If you have the animal adequately restrained, his head movements are usually not extreme.

Leather halters retain shape and are usually easier to put over a llama's nose than other types of halters. We sometimes put the halter on and off several times to get the animal used to putting their nose into a halter. Additionally, leather halters are usually made so that the nose piece adjusts along with the rest of the halter as you pull the strap through the buckle.

Surprisingly, the untrained males are usually easily trained to lead in two or three sessions. If you talk to them in a kind

BASIC TRAINING FOR THE ADULT

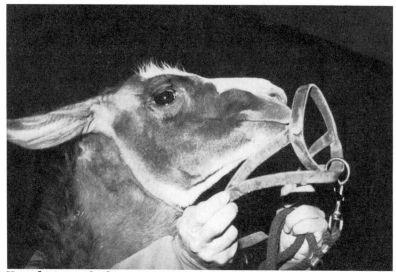

We place a halter on an adult llama in early stages of training. The halter is slowly moved onto the face.

soothing voice, they respond quickly to your wishes and praises.

Training an adult to load in a trailer or truck once he has learned to lead is an interesting process. Usually if you can get them into the trailer one time, you have it made! But the first time may require some effort. Sometimes a second person has to pick up each front leg up and place it in the trailer while one person stands in the trailer and pulls. Then shove on their rear and in they go. You may even have to put their front legs in the trailer for the second loading. But after doing this only once or twice, its as if they turn on a loading switch in their brain, and they jump right in on the next try.

One of the fighting methods of male llamas is to bite each other on the legs. It is not natural for them to allow their legs to be picked up. Desensitizing the legs of an adult male can be easy or more difficult, depending upon the individual llama. Adult males tend to try to lie down more than a weanling does when you touch their legs. If we have an animal that kneels down on his front legs when we touch him, we sometimes loop the lead to something higher than the animal so that he cannot lie down. If he gets into trouble, you can quickly give him some more lead. Sometimes a quick, hard pat at several places is better than sliding your hand around or rubbing the animal. If you are tall enough, you can prevent an animal from lying down by leaning across the animal's back and putting some weight on him. The usual reaction is for a llama to push in the opposite direction, which would be up so this prevents him from lying down.

When he gets used to you touching his legs, you may attempt to pick up one of the legs and then quickly return it to its normal position. It might take several sessions before he will allow you to lift a leg. I usually work on front and back legs during the same session. Remember with the back legs, he might try to lie down or to kick. Keep your body away from those back legs. By positioning my body along side the middle of his body, I am tall enough that I can reach back and touch the back legs. Some use a training aid, such as a wooden dowel, or stiff piece of leather to reach back and touch the back legs. The remainder of the training is similar to that given in previous chapters for weanlings.

Training an adult can be very quickly done. Adult llamas are smart and learn very quickly. With a kind voice and gentle training, they learn quickly and like to please.

CHAPTER 8

TRAINING - PACKING WITH LLAMAS

About the simplest thing to do in your training program is to teach your llama to haul a saddle and panniers. It is much easier than teaching a new weanling to lead. It is almost in the category of teaching a mature stud to be interested in romancing a female or getting a baby interested in nursing. It almost comes naturally to the llamas. This is particularly true if your llama has learned the previous lessons well. An animal whose ancestors have been used continuously for packing for 4,500 to 6,000 years isn't exactly a newcomer when it comes to packing. So we will discuss putting on the pack later.

There are, however, several things you can do that will make your young student feel a little more at ease when he goes on the trail. Several obstacles may worry the inexperienced llama. Among these are creeks, bridges, low over-hanging limbs, tree trunks leaning across the trail or logs on the trail.

For the inexperienced llama, a creek, river or bog may appear as a formidable obstacle and perhaps a life threatening situation that is to be avoided at all cost. I remember when we took an inexperienced llama to the mountains for the first time that was not acquainted with water. He wasn't carrying a pack and was tied to the saddle of an experienced llama. We crossed a tiny two-foot stream of water. He held back until he could no longer avoid being pulled forward then he jumped with all his might easily clearing the water by six to ten feet, landing beside the llama he was following. This type of behavior is dangerous to the jumping llama and can even

LLAMAS ARE THE ULTIMATE

A portion of our "formal training" consists of the class lecture. Five large llamas listen intently before being saddled for the first time. Two were less than two-years old and the others were two.

injure you if you happen to be in his landing strip. A jump like this can break a leg if he lands on an underwater boulder.

The fear of crossing water can be conquered in one or two easy lessons. Sage Creek which runs through our new farm provides the media for the solution. Prior to moving to the farm, we took our llamas to an irrigation canal.

WARNING: This procedure requires you to get wet! If you don't like to get wet, and believe me, the water in canals in this area can be cold, you can wear waders. This is one procedure that works best by you setting the example. It may also help if you lead an experienced llama into the water

Snake River Tuxedo relaxes beside me after only a few minutes of being in the water for the first time.

LLAMAS ARE THE ULTIMATE   41

while your new student watches. However, we don't think that putting an experienced llama in the water is necessary. You may need help in getting the llama anywhere near the water. Once you are close, calmly walk into the water. You may have to force the llama into the water by pulling on the lead while your assistant pushes on the llama's rear. We are careful to choose a spot that is too wide for the llama to jump and does not have rocks that could cause injuries.

At this point, llamas usually allow themselves to be pushed into the water, they may even jump a short distance into the water. We had one youngster, that jumped as far and high as he could, so don't be surprised if your llama comes by. Use the lead to keep the animal in the water. Talk to him in a calm, soothing voice. He may practically climb onto you as he sees you as a form of security. After a few moments, walk a few steps up, down or across stream. Then lead your llama to the edge of the stream. He will probably eagerly jump out of the water. Immediately pull him in again. Remember to talk to your llama. Within a few tries, he will step off into the water on his own.

When we took our llamas into the canals, the water was about two and one/half to three-feet deep and was up about half-way on their sides. Seldom will they encounter water this deep, so this prepares them for just about any water condition. We have taken llamas back to the canal after one lesson, and they walk right into the water following me on a loose lead. The fear of water is usually overcome in one lesson.

With just one lesson, llamas usually do fine on the trail when they come to water. They may hesitate a little on the first few crossings, but they will generally enter the water without too much difficulty. We've found that if you try to jump over a stream of water, your llamas will often try to jump the water also. You may be better off wading the stream if your llamas are tied together in a string. A jumping llama causes the lead tied to the back of the first llama to jerk the next llama.

Training llamas is very similar to training dogs. You want to remain in control of your emotions at all time. If you lose your temper, it would be better for you to discontinue the lesson and try again another time. Llamas do not respond well to harsh treatment. It is usually unnecessary and harmful to his learning experience. Try to talk softly and reassure your llama while working with him. When he does something right, praise him! Again, several short lessons may be better than a few long ones, particularly if you are working with a young llama.

Another natural barrier is the low, over-hanging limb or tree across the trail which requires you and the llama to bend down to navigate under it. When encountering this condition on the trail, most llamas who have not experienced going under a limb or tree will balk. If you don't have a low limb parallel to the ground, there are a number of man-made structures, usually located nearby that will serve to train your llama. A bridge or foot board across a creek or canal that permits you to pull your llama under it or a large cement or metal culvert to go through will work if they require your llama to at least hold his neck and head down to pass under or through them. Encountering these conditions a few times on your walk with your llama will be all that is necessary to train him to go under tree limbs or trunks on the trail.

Another natural barrier that might impede your llama on the trail are tree trunks lying flat on the ground. The teaching technique is quite simple and obvious. Again, it is something that can be learned on the trail. If you're going to teach your animal on the trail, it is best to have him separated from the

LLAMAS ARE THE ULTIMATE   43

A young llama is taught to jump a log on our farm. Later he will learn to step over or to jump only as high as necessary.

rest of the pack string. However, it makes the animal more comfortable and manageable if you teach him to jump logs prior to going into the mountains.

When we were disposing of all the old machinery, limbs, trees and junk at our new farm, Cherrie had the equipment operators leave two large old tree trunks lying near the creek. They are of different diameters and both require a llama to jump to get over them. The smaller diameter trunk is used for beginning training or for younger llamas. Objects such as downed trees and culverts can probably be found in your neighborhood. Of course, the object is to start your llama jumping a small log and proceed to larger ones.

When we camp with llamas, we tether them out on a 25-foot cord so that they can eat. If you plan to stay nearby, you may want to tether your animal for the first time when you are

TRAINING - PACKING WITH LLAMAS

A llama snapped to a 25-foot tether has plenty of room to graze.

LLAMAS ARE THE ULTIMATE

trekking. The nylon webbing or rope attaches to the llama's halter and ties to a tree, screw-in-stake or other object. WARNING: A tether that has a weak link (for example one that is lightly sewn on the halter end) that will break if the animal panics is well worth considering; although a rare occurrence, llamas have been known to break their neck on tethers when they have panicked. During the first night or two on the tether cord, he will tangle the cord around his legs and will need your assistance.

You may prefer to tie him out on the tether a few times before he leaves your ranch. However, don't tie him in a corral or field where other llamas are present. Both geldings and studs like to jostle each other, and they may take advantage of one being tied.

Llamas are smart and can learn in a very few tries to watch the cord and not become tangled in it. However, that is not to say that they will not tangle the cord around trees and bushes where they are grazing. Generally if they tangle the cord around shrubs or limbs and get in trouble, they will simply lie down and wait for you to come and get them loose so they can continue their grazing.

Teaching your llama to actually haul a pack around is one of the simplest procedures and one of the easiest to learn. There is no need to purchase any type of practice saddle/panniers as it is completely unnecessary. The pack you plan to use in the mountains will work great for your training.

Let's assume that you have trained your llama; he leads well, his legs and body have been desensitized since an early age. At this stage (if not before on an older male), we teach them to haul a saddle and panniers. Two years of age is an ideal time to teach this to llamas. We have placed saddle and panniers for the first time on llamas that were up to eight-years old.

We tie the llama and let him see the saddle pad as we begin to put it on his back. Cherrie thinks that putting the packs and saddles near the llama for a few days gets him used to the equipment and makes the next step easier. After placing the pad, we immediately put on the saddle using only the two cinches to secure the saddle to the llama. Be careful not to catch any wool or hair as you securely tighten the cinches. If you have trained your llama adequately, generally he will hardly move during this procedure. We like to do this step together as a couple. If you are doing it alone, tie the llama fairly short using a quick release knot or a quick release snap and use your body to keep the llama from jumping about. The first strap goes immediately behind the front legs. This cinch needs to be very tight; it does most of the work of keeping the pack on your animal. The rear cinch generally goes at a slight angle backwards on many packs. It rides about half way between the penal sheath and a point directly beneath the cinch attachment to the saddle extended under the llama. The type saddle you use may be different so follow the manufacturer's instructions for your particular saddle.

If your llama is not well trained and is inexperienced at being handled, he may want to lie down during this procedure. If he does, tie his lead up above his head so that he cannot lie down. Its a good idea to tie him so you can quickly untie him if he gets in trouble. If he has a problem with the cinches or straps touching his legs, you can tie a piece of twine on the end of the cinch strap and have your partner pitch the twine underneath the llama.

Next put on the empty panniers which are easily placed over the crossbars on the saddles we manufacture. We then untie

LLAMAS ARE THE ULTIMATE

A saddle is put on Hershey for the first time. Make sure that you don't catch some of the wool in the cinches.

TRAINING - PACKING WITH LLAMAS 48

the animal and lead him about. Don't be too disappointed if your llama hardly reacts. Llamas are very different in this regard. Some may pitch and jump for a few minutes before they settle down and some hardly react, while others will be somewhere between these two extremes. Their degree of initial reaction has no correlation to their behavior as pack animals. Usually within a few minutes, they will calmly follow you around as you lead them. This would be a good time to take your llama for a walk, going over some of the obstacles mentioned above.

Although Jasper was nervous and pulled back on the lead a little, these were his only reactions to having the saddle and panniers on for the first time.

The llamas quickly adapt to the saddles and panniers. On several occasions, we have taken youngsters to the mountains with packs after having the saddles on only once or twice.

Sometimes a llama will quickly adjust to the pack, but will momentarily panic when a willow or other brush scrapes the side of the pack. To avoid a llama trying to self unload the panniers on the trail, go by a lilac or some other bush in your yard a few times and let the limbs brush against the packs. Don't be too surprised if his reaction to the noise of the branch against the pannier causes even more of a reaction than putting the saddle and panniers on for the first time.

One way for your green llama to learn to pack quickly is to tie him behind an experienced pack animal. Llamas usually adapt to packing in the mountains so easily that I consider them veterans after only one or two treks, especially if they have mastered the basic training before being packed. We can pack all sorts of game like javelina, elk, deer and moose on young llamas that have packed only once or a few times. To the llamas, it is just another load.

To those of you who might think that teaching a llama to pack is a complicated procedure which might even have to be done by a professional trainer, I apologize. Training a young llama to pack is usually a rewarding task, which most people enjoy doing themselves. Particularly, if you plan to be the one packing with the llamas, I recommend that you do the training. You will then feel at ease with your animal on the trail the first time you go out with him. If you are a newcomer to owning llamas, there is nothing better than actually getting some experience with your llama. You'll learn a lot about llamas in a short time. It will increase your confidence a great deal and will be an asset to your llama management program.

CHAPTER 9

WHAT AGE CAN YOU START PACKING YOUR LLAMA?

I know a couple who start taking young llamas packing with their adults when they are eight-months old. They don't pack anything on them, but when they do begin packing them, they are seasoned veterans. Often they tie their adults together in a pack string and let the youngster run free. As they trek along, the youngster occasionally stops to graze but runs to catch up if the others get very far ahead of him. Although this method works for them, I recommend that you lead or tie your young llama to another llama if you take one along on the trail.

We normally do not pack our males until they are between two and three years of age. However, if you desire to pack your yearling llama, it will not hurt him at all if properly done. I'd suggest loads like sleeping bags and other lightweight equipment. Keep the load below 10% of his body weight. It is always a special thrill the first time a new llama packs. If it's your first llama, it will be even more satisfying to see how easily he adapts to packing. It's easy to make the mistake of assuming that you had a lot to do with him packing so eagerly and easily.

Don't overload or overwork your llama the first time out, particularly if the weather is hot. Try to build up his strength and endurance gradually and make it fun for both of you. With experienced mature packers, we often expect more out of them on their first trip of the year, but be careful with the

LLAMAS ARE THE ULTIMATE

Castles at 22 months and 340 pounds and Buffey at 27 months and 370 pounds dwarf Cherrie on their first packing trip. They will be big boys when they reach maturity.

younger and inexperienced animals. You will want the first packing experience to be pleasant for your youngster. I don't always follow this advice. Recently we took two inexperienced llamas, which had never packed before, with us on a 58-mile trek. Amazingly, the 23-month old and 25-months old llamas carried up to 50 pounds and traveled the entire distance like veteran llamas.

At the other extreme, llamas can be started to pack at most any age. We put one in our pack string when he was nine-years old, and he had never packed before. He's a big tall llama that weighs over 400 pounds. He didn't get tired very easily and was one of our better packers. Francis Greth-Peto

WHAT AGE CAN YOU START PACKING LLAMAS

of California has told about putting a llama in her pack string when he was 15 and packing commercially with him until he was 27-years old. So don't be afraid to purchase a mature adult for packing. Adults learn very quickly, and it is likely that llamas can pack for the weekend and vacation packer until they are into their 20's.

CHAPTER 10

HOW MUCH WILL THEY CARRY?

The amount of weight recommended for hauling on llamas depends upon their weight and age (Table 1). If packing over a long distance, walking upward over a significant elevation gradient, or packing for more than one day at a time, we don't usually load our animals to the maximum weight, particularly if they are not in the best of shape. It will depend upon the conditioning of your llama; however, one in only fair condition should be able to easily carry a maximum load four to ten miles without any difficulty. During hunting seasons, we often expect our llamas to carry 33% of their body weight. We have loaded them with even more than that. Although we have exceeded the recommendation, I don't advise it be done often.

TABLE 1. Recommended packing weights as percent of body weight for different-aged llamas.

AGE	AVERAGE LOAD	MAXIMUM LOAD
>one	10%	10%
two-three	<20%	<20%
four	20-25%	33%

Last fall, a friend of mine took four llamas five miles to haul out a raghorn bull (four point antlers on a side) elk for his brother. While they were in the area, a spike bull wandered

HOW MUCH WILL THEY CARRY? 54

Himalaya at 365 pounds hauls out a two-point mule deer for Cherrie. The panniers weighed 135 pounds, slightly over the recommended weight for Himalaya.

by and my friend shot him. Now they had two elk to haul out on four llamas, and they were five miles from their truck. They skinned and quartered the elk. They put the bigger bull on the two adults and the spike on the two-year old llamas. He later weighed the loads and found that the two adults who weighed about 310 and 340 pounds had 172 and 178 pounds of elk, respectively, and the two-year old llamas were carrying 135 pounds each. Obviously, these animals were carrying much more than the recommended 33% maximum. Some llamas are capable of carrying tremendous loads. My friend had not taken his panniers along so he tied the quarters directly on the saddle. With panniers, he could have boned the meat and made the weights more reasonable.

Sometimes circumstances may make you think that you have to put more weight on a llama than what you planned. However, putting too much weight on them does have risks associated with the overloads as you may injure your llama. I strongly recommend that you do not pack your llamas with the proportions of body weight (52-55%) as my friend did.

If you have a set of panniers with a heavier load, you can change the load between various llamas so the same animal doesn't always have to carry the heaviest load.

One Easter weekend, we were trekking with a group of llamas in southern Idaho. One of the llamas was a younger animal out for his first trip. We crossed a river canyon but could not find another crossing through the steep canyon. We eventually packed 25 miles to reach another road. Although it was the first trek of the year, all the llamas, including the youngster made it okay. As previously mentioned, we recently took two llamas 58 miles on their first trek.

A few times, we have packed 25 miles with llamas in a weekend. We know of others who have packed up to 18-24 miles in a day with llamas. However, llamas should be in good shape and loaded moderately to go these distances. You certainly would not want to push llamas in only fair condition this hard for more than a day or two. However, at a race held yearly in Mancos, Colorado, llamas carrying 50 pounds each race over a 16-mile course that reaches up to the 12,300-feet elevation. These llamas race the course with several handlers. The winning times during the first year's race was just over four hours. Times will likely improve each year. Llamas in good shape can cover lots of ground in a hurry.

A good mature packer knows what is expected, and he will get the job done. I have read and heard that llamas can't be

HOW MUCH WILL THEY CARRY?

loaded too much or be forced to walk too much because they will just lie down if they are overloaded. Don't you believe it! A young or inexperienced llama might lie down on you, and we have previously discussed this possibility. However, many experienced veterans have a lot of pride and heart; most will pack until they almost drop before they lie down. However, you have a responsibility to these animals to use them wisely. Just because they will display amazing willingness to do that first job of the year, no matter how hard, doesn't mean that it is the best way to treat your animals. Sometimes circumstances, as in the two cases previously discussed, force you into going further or loading them with more than you would like.

In my opinion, a llama should not be packed, at least for long distances, with a full maximum weight until he is three and a half to four-years old. Most llamas are at full height and some of them are near mature weight earlier; however, their bones and muscles need the strength that an additional six months will give them.

For most summer packing trips, we load our llamas with 75 to 90 pounds each. To give you an indication of body weight packing this load, we normally pack with llamas weighing from 360 to 430 pounds.

You should know the weight of your llama before packing with him unless you are loading him with light weights that you are absolutely sure are no where near his maximum weight for his age. WARNING: Weight estimates of llamas even by experienced llama breeders are often no where near correct; don't rely on estimates.

If you don't have a scale, you can usually weigh an animal on a commercial scale for $1 to $3. Any livestock sales company and many meat packing companies have scales. Commercial scales can also be found at hay, coal and other commercial outlets. If the scales you use are designed for hay trucks or trailers, don't weigh your llama in your rig and then weigh the rig again without the llama. The llama weight will be much more accurate if you remove your llama and weigh him by himself. Livestock scales will likely be more accurate than the truck scales.

You may have purchased a scale specifically for weighing llamas, so you are getting correct weights. Right? Not necessarily, some of the scales that you see at various llama functions and in catalogs may not weigh accurately or consistency. At a llama function, a friend of mine weighed on one two days in a row and his weight differed by 35 pounds. The difference in his weight on two different brands of scales was 22 pounds. Other llama breeders have told me about purchasing scales that were so inconsistent that they would not weigh within 20 pounds of the accurate weight. You'll need to test your scales both for accuracy and consistency. You just can't assume your scales are correct just because they cost you a bundle. If you have access to a medical scales or you have calibrated your bathroom scales with that of a medical balance, then you can be a test for your scales. It is better yet, when several know their "correct weight" and can weigh in combinations to test your scales. There are certainly better ways to test your scales, but this method will be surprisingly effective. The weight of a llama varies by season and with daily feeding, drinking and elimination patterns. Don't be surprised if your llama varies by 15 to 20 pounds. Also a week of hard packing can have considerable effect on the weight of a llama.

HOW MUCH WILL THEY CARRY? 58

By loading your llamas with the proper weight for their age and physical condition, your llama will provide years of packing enjoyment.

CHAPTER 11

ON THE TRAIL

You have your llama trained, and he is ready for the first trip into the mountains. He has received his basic training and has carried a saddle and panniers at least once. Now it's time to reap the rewards and for both of you to enjoy the best period in your lives!

Usually you will have to transport your animal to the trail head in a van, truck or trailer. You can either tie him or let him freely move about. If the animals to be transported are fully compatible, it is better if none of them are tied. If you tie him be sure to allow him enough lead to lie down. However, if there are more than one llama present in the trailer or truck, make sure the free lead isn't so long that another llama can come between a llama and its tie location. This can force a llama to ride with his head stretched out. Also, if your lead is too long, another llama can get the lead wrapped around its neck.

It is recommended that you get your animals in condition before trekking long distances. We used to pack loads of wood or other heavy materials around on our llamas several times before trekking each year in the mountains. However, we no longer have the time to do this. Now, we usually try to make some shorter treks the first time or two we go out to get the llamas in shape for longer treks. However, in general, experienced llamas will do the job, in shape or not, even on a long, hard trek.

When you reach the trail head, he should allow you to load saddles and panniers while he is tied to the side of the trailer or truck. Perhaps you didn't have as much time to give him his basic training as you would have liked. Maybe he is a little apprehensive as well. For whatever the reason, he might decide that he will lie down as you try to reach under him. Just tie the lead up high so he can't lay down or move about. Usually, they saddle very easily if they have received the previous training.

Each time you put a saddle on your llama, you should check the wool to make sure the wool is free of sticks or other objects that would cause discomfort for the animal when he is saddled. Also check where the straps fit under the animal. Usually, I run my hand over the area where the pack or saddle will ride and then underneath before putting the saddle on. Many packers use plastic curry brushes to remove debris from the top of the wool.

In the spring, some owners like to remove all the loose wool from their animals by brushing them with a wire-bristled brush, commonly sold at pet or sheep supply stores. This is a time consuming job, and a llama usually grows impatient before the job is done. Often it is better to do the job in several sessions. Other owners prefer to shear part of the wool with hand-operated sheep shearers. If your animal is light to medium wooled and you pack in dryer climates, you will likely not need to remove any of the wool. If you blow the wool with a leaf blower or commercial livestock blower (available at llama supply dealers) before grooming, the debris and dust can be removed from the wool. It's usually essential to blow your animal if you intend for the wool to be sold or used in spinning. The blower also makes the animal look better and cleaner for packing.

From what side do you load a llama with saddle and panniers? Both, it should not make any difference. We sometimes assist each other by handing the cinch straps under the animal to the person tightening them. Now is the time to appreciate the training you have done on desensitizing the legs of the llama. He should stand motionless when the straps touch his legs or you touch them in reaching for the cinch straps.

Even from the beginning, he should easily follow as you start up the trial. However, you may find that after aways, he starts to slow down, and you have to continually jerk him forward. This would not be too common for beginning llamas. However, it sometimes happens. The first time we packed with Bandit, I had to continually jerk him to get him to move at a reasonable speed. My arms were tired when we reached the lake. On the next trip, he led like a dream, and he has been that way ever since.

You should not allow your llama to get into the habit of poking along. You want to set the speed, not him. It's a good idea to vary your speed so he will get used to moving at any speed. Also make sure that he follows you, and does not try to be the leader. If he has a tendency to want to pass, reach out and touch him on the face or jerk him back in line behind you. Another technique some folks use is to twirl the end of the lead in front of the llama. However, make absolutely sure that you don't unintentionally hit his eyes.

Some llamas on their first trip or two will decide that they may be getting tired. When he is tired, he is accustomed to resting, usually by lying down. Even an adult on his first pack trip may decide that its time to lie down. Don't allow him to remain lying down, even if he is tired. When you have a llama lie down, do some combinations of hollering and running at him, waving your arms or jerking on the lead. This usually gets him up. If it doesn't, try rolling him over or pulling on his hind legs. (This assumes that he is just tired and not undergoing heat stress or other physical problems). Once he is up, you can continue on for a short distance. If the llamas in your group are obviously tired, then stop and allow him to rest. This way he learns that he can't just lie down anywhere, and you pick the rest periods, not him. This will usually only happen a time or two if at all. If it is corrected early, it will usually not happen again.

One of our llamas had a sore hip one year while on the trail with another person. He was obviously hurting. On the trip back to the trail head, whenever she stopped for a rest, the llama immediately laid down, but when she got up to go, he immediately joined her even while obviously hurting. Llamas have lots of heart and pride!

You can usually tell when your llamas need rest. If you're in reasonably good condition and you are having difficulty and would enjoy a break, so would your llama. If you have been steadily climbing for more than an hour, you should break for a few minutes. If a llama begins to breathe hard, and you can hear them breathe even after stopping for a few minutes or rapid breathing can be seen, stop and give them a rest. A few breaks along the way will actually help them continue a faster pace when they are walking. This will also help avoid heat stress problems in the summer.

Llamas like to nibble a few plants as they walk along the trail. I usually don't mind so long as it does not slow the pace. Llamas learn to take a few bites as they walk along almost without breaking stride. However, don't allow your llama to develop a habit of eating anytime he wants and delaying you in the process. This can develop into a real problem later, especially when a smaller or inexperienced person is trying to lead the llama down the trail. If you are trekking greater distances, it's a good idea to take your breaks where there is plenty of forage and allow your llama to feed during the break. This will help satisfy the llama and help prevent him from wanting to stop along the trail. Making sure that they are adequately fed at night or before the trip will also keep them going. A quick jerk on the lead is usually all that is necessary to keep him from stopping and eating along the trail. A llama that wants to stop on the trail and eat can be put into a pack string to help reduce this tendency. The llama in front of him will jerk him forward when he stops to eat.

Although you may have had your llama in water once or twice, don't be surprised if he balks at the first crossing. Usually the problem is not severe, but you may have to encourage him to step into the water. Usually one person can easily convince him to move into the water, if he has been in water previously. Tug on his lead rope and talk to him. At least he won't be terrified of the water if you have done the water training during your basic training program. Cherrie thinks that if you try to jump over the water, the inexperienced llamas will jump too. They will have a tendency to step into the water if they see you walk through it.

When you cross a tiny stream or small gully, your llama will want to jump it. Even veteran packers prefer to jump these small obstacles. This presents a frustrating and interesting situation when you have llamas tied in a string. As one jumps, the lead between the two jerks the saddle on one backward momentarily while the following llama's head and neck get a good jerk forward. I wish that I could tell you a method to teach your llamas to cross these places so they would always walk across them even in a pack string. But alas, I cannot. Even the best trained llamas will jump these small obstacles when they can, particularly if they are in a pack string.

A llama with proper basic training on the ranch will usually experience no difficulty in crossing streams on the trail.

LLAMAS ARE THE ULTIMATE   65

A very good method of training a green llama is to tie him behind another. Usually, a beginning llama has no objection to being tied to another llama, particularly if he is familiar with the other llama. Usually, it is not a good idea to tie another llama onto a first time packer. They do much better if they are allowed to follow for a few miles or a trip or two. Depending upon the individual llama, they can quickly adjust to being in other positions in the pack string, once they have had the experience of following. It varies between llamas, but some of them can be switched to other places in the pack string after having only a few hours experience in following.

Even beginning llamas, when packed with veterans, can go places that will amaze you. This extremely steep decline along with wide snow fields scared me, but the llamas did the job without problems.

Llamas generally do not like to cross muddy or boggy areas. Sometimes these conditions do not exist on or near our home farms or ranches. So when you see these areas with your new packer, be prepared for a bit of a problem. He may actually refuse to go into them, or he will attempt to jump as far as he can. In time, they will learn to trust your judgement and if you step off into a boggy area then they will also. I don't guarantee that they will ever like it, but at least you can get them across without too much trouble.

Once we took our llamas on a trial that led across a three-quarters mile boulder field. The boulders had spaces between them that could have seriously injured llamas had they stepped in them. We moved slowly across the boulders, giving the llamas plenty of lead so that they could see and pick their way across the boulders. They easily traversed the boulders both coming and going.

Although it is rare, a few llamas do panic for one reason or another. Sometimes its a new to packing, older llama that has never been packing or a youngster that is going out for the first time. But it could happen with any animal, even an experienced one. So be careful how you hold the lead, even with the most trustworthy llama you own. Never wrap it around your hand or any other part of your body or belt. Always hold it in such a way that you can instantly drop it. If you need to have your hands free, you may want to tuck a loop of the lead in your belt so it can easily be pulled out if the animal were to panic.

Although your llama may have been staked out a few times at home, it's a good idea to keep a close eye on him the first night or two.

Your llama will do much better if he is packed early in the morning when it is cool. In western mountains, heat and humidity are seldom a serious problem. However, in the east and south, you may not want to pack at all in the middle of the day and perhaps not at all on certain hot and humid days.

Cherrie and Brutus, on his first pack trip, in a three-quarters mile wide boulder field. The llamas were impressive as they avoided all the holes between the boulders.

ON THE TRAIL 68

Some caution is necessary when crossing snow fields in the middle of the summer. Llamas sometime "fall through."

To your surprise, you may find that on the first day out, your llama refuses to drink any water. Perhaps you've heard the old saying, "You can lead a llama to water, but you can't make him drink." This may not be Biblical, original or even quoted correctly, but it is gospel. Don't be too surprised if on the second or even on the third day, the same thing happens. On the other hand, your llama might drink more than once each day, if it is particularly warm. One spring, Mike Tupper and I trekked in the Frank Church River Of No Return

Wilderness in central Idaho. One of the llamas with us did not drink until the fourth day. We offered him water several times each day; he just wasn't thirsty.

I have read that some llamas refuse to drink from swift streams and, therefore, a bucket of some sort is a good item to bring along. I don't carry a bucket for llamas, and I don't believe it's cruel to insist that they either drink from the creeks or go without water. If they get thirsty, they will drink from a swift stream.

Usually after just one trip, your llama will act like a veteran packer. Once on the trail for that first time, you'll find that if he has been trained properly at home, he'll be calm and cool on the trail and nothing much will bother or excite him. Once a llama has been out a few times, he'll perform equally well even on the first trip of a new packing year.

CHAPTER 12

PACKING WITH A PACK STRING

I love to pack with a pack string. I do such childish things as step out of the trail onto a meadow and stumble around while looking back to see each male step out of the trail at the same place. We usually have from two to six animals in a string; however, we have had six or more on several occasions. Three to five animals is an optimum size and can be managed anywhere. One time, while hunting in Wyoming, I put seven (most had very limited experience) in a pack string to cross an open meadow with scattered trees and a poorly defined trail. It was interesting and entertaining, but it wasn't very efficient. However, with the right conditions and experienced llamas, pack strings larger than five llamas can work very well together. Recently, we had nine animals in one pack string for a half day and covered about five miles without any problems.

Unlike other common pack stock in North America, both studs and geldings can be packed. We always pack with studs and geldings in the same pack string. When we leave the farm, the mature studs are tied in the trailer with enough lead that they can easily lie down. Our geldings are not tied and are free to move about in the trailer. Studs may be aggressive at this point toward the others or the geldings are frightened of them so the studs should probably be restrained. By the time they have traveled to the trail head, both studs and geldings have settled down. We then alternate stud and gelding in the pack string. Usually all are well behaved and spitting or threats are rare. They have forgotten about the females and know they are in a working situation. After a day or two, we usually can pack the studs next to one another in the string and they can travel unrestrained in the trailer. Within a day or so, both studs and geldings will miss each other's company if separated on the trail. These same studs might be mortal enemies at home around the farm.

If you are packing with several llamas, whether in a string or not, the llamas will be much better packing companions if they have been boarded at the same location, and preferable the geldings have been in the same field or corrals. If you have ever brought a new animal to your farm and suddenly introduced him into the male pasture, you know about the running and chasing that usually results for a few hours. Similarly, packing with llamas from more than one herd results in a lot of rear end smelling, threats and other problems, and the potential for some spitting is increased.

If you have a commercial operation or let a variety of people handle your llamas, you probably should consider packing only with geldings. However, if only a few people go trekking with you, and you would like to pack your studs along with the geldings, it can be easily done, and I would encourage you to do it. The only precaution is that adult studs not be packed next to one another unless you have allowed a day or so to go by after leaving the females. Clients might not remember this simple precaution. In addition it will provide some valuable exercise and conditioning for a stud and will likely make him healthier and more valuable to you. This will also increase your packing pleasure and will reduce your need for additional geldings.

If you pack with a string of llamas, it is usually better to put the green llama at the rear of the string. Depending upon the individual llama, you may want to tie him on to a single llama, making a pack string of two. A couple of falls ago on an elk hunt, I tied a new llama as the last one in a four packer string. He followed along without any problems and one might have thought he was a seasoned veteran. However, he had only had a saddle on twice, once a month before the hunt and once the year before. He was sure gawking at the surroundings and occasionally wondered out of the trail because of it. This is the normal behavior for a first time llama; most adapt to packing in a string instantaneously. It usually is a simple thing to teach llamas to pack in a string if they have been trained adequately at home, and they are compatible with one another. If they will singly follow you, they will follow other llamas just as well. If your green llama has mastered all the previously discussed basic training, he'll do fine the first time. After a few hours out, he can be switched to another position in the pack string. If your llama is poorly trained and has not mastered the other lessons, he probably needs to be led by himself the first time out. Also he may not do great if he is a visitor to the rest of the animals.

Usually a llama is tied with the lead to the back of the saddle or pannier of the llama in front of it. With saw buck saddles, a loop can be made in the lead rope and loped over the back cross bars of the metal frame of the saddle. Then the loop can quickly be lifted off the saddle in case of trouble on the trail. On ours, we installed a ring to which three low-tensile pieces of twine can be tied. The lead can then be tied, preferably with a quick release knot, to one or more of these twine loops. If it is a dangerous trail where a llama might fall, then one loop can be used. Then an animal can easily break away from the one in front of it if one animal falls. A quick release knot can also be used to tie the loop or to tie him to another part of the saddle.

I prefer to vary the positions of the packers when they are first becoming accustomed to packing in the string. I usually start

LLAMAS ARE THE ULTIMATE 73

My little friend, 60-pound Kara Clark, easily handles nine big llamas in a pack string in the Wind River Range of Wyoming. This string packed together for a half day without a hint of trouble.

with a stud in front of the string. But over the course of the trek, most of the llamas, and especially the new ones have the opportunity to pack in all positions. This will prevent a llama from wanting to pack in only one position. However, you may find that a particular llama that you own might do better in a particular position in the string.

If you are apprehensive about packing with a string for the first time, pick out a trail which is bounded closely with dense trees or rocks. In these situations, the llamas have no choice but to follow one another down the trail. For the first day or so a llama packs in a string, the lead rope from the llama behind him may bother him quite a bit when it swings under

PACKING WITH A PACK STRING 74

his tail. Generally they get used to the lead moving around their tail fairly quickly. But you may see a llama suddenly threatened the one behind him when it happens the first few times.

Some people have a favorite way of hooking their llamas together when they start out with their llamas. Some start with the last llama and attach him to the next one, etc. until they arrive at the lead llama. Others do it in the reverse order. I don't have a standard method that I use each time. However, we have found that if the load on each llama permits easy access to the rear cross bars of the saddle, the loop of the end of a lead can be quickly thrown over the rear horns of the saddle on the llama in front of him. All the llamas can then be attached to one another very quickly.

Kasi Clark leads five llamas packed with over 500 pounds of supplies. Large llamas are able to haul big loads when necessary.

LLAMAS ARE THE ULTIMATE 75

With experienced packers, a string of llamas follows so smoothly that it will seem like you are leading only one good packer. However, you should look behind you every few minutes to check to make sure that nothing is wrong with any of their packs or if one of them is having difficulties. This is especially needed when you are in rough terrain or going over obstacles. Remember after you and the lead llama have crossed an obstacle, continue to go slow allowing all of the llamas in the pack string to be able to cross the obstacle at a slow pace.

Pack strings are particularly efficient during hunting seasons or in any situation where one or two people need to pack large amounts of gear into a remote camp. Packing out moose or elk usually require a pack string.

Sometimes llamas in a pack string find it difficult to drink water when they come to a creek or river. They may not have adequate opportunity to drink because the llama in front of them has pulled them out of position to drink or they may feel uneasy about drinking with their traveling companions all around them. So occasionally you may need to give each individual llama the opportunity to drink. This is easily done and might only require you to have a llama in each hand or to tie one or two while others drink. Since you only have to do this once or twice during the day, it isn't much of a hassle.

When we reach our destination, we tie the llama we are leading to a tree or shrub and then detach the rest and tie each to a separate tree. Then we can proceed to unload each animal, either where he is tied or at a common unloading area. Camping with several llamas is no different than packing with one or two. It's a good idea to have a little grain with you to give them as a treat and to make sure they get adequate feed. The time of the year, available forage and other factors will

PACKING WITH A PACK STRING 76

Pack strings enable a person to take along many conveniences and pack large amounts of supplies necessary for extended periods in the wilderness.

need to be considered in how much grain to pack with you. Sometimes the best and most scenic place for you to enjoy may not be the best place for llama grazing. So if you are unfamiliar with the area into which you are packing, you may want to take a little extra grain.

Some people let some of their llamas free graze while the natural leaders are tied or at least part of the herd is tied. Usually the free grazers will stay near the others. A small bell tied onto them will make them easier to find in the mornings. Be careful about letting all of them graze free or you may find that they have good memories and can make all the proper turns to return to the truck without your assistance, leaving you and the equipment alone in the mountains. We usually tie all our llamas at night.

When you are trekking with a string don't be surprised if one llama defecating starts a chain reaction for all of them.

A young girl walks through a field of flowers with two first-time packers and three veteran packers. Packing with a llama pack string is about the ultimate in packing with llamas.

We used to pack without the use of a breast strap or a britchin strap on our saddles. With the type saddles we make, it wasn't necessary to have the additional straps when leading only one animal. However, when packing with a string, there are always circumstances where a breast strap is beneficial. No matter if we are packing with experienced or inexperienced animals, when we come to a small stream or small ravine, llamas like to jump them. Of course when they jump, it snaps the lead on the llama attached to them. This causes the saddle to shift toward the rear, and the breast strap prevents backward movement of the saddle. We don't use the rear or britchin strap with our saddles, which I think makes a neater set up (we have them as an option) and really isn't necessary. However, they are an integral part of many saddles.

Often a slow llama can be made to go faster by putting him in a pack string. When a llama in front of him snaps his head forward a few times, he gets the message that it is more enjoyable to keep up a steady pace with the rest of the string. Some llamas are excellent teachers of other llamas. Some will threatened the llama behind them if they pull backward or try to pass on a narrow trail.

Cooperation and congeniality in a pack string is almost a natural for llamas. Therefore, trekking with a string of llamas on a trail through a meadow carpeted with flowers on the way to a scenic high mountain lake is about the ultimate in owning llamas.

CHAPTER 13

TRAIL AND CAMP ETIQUETTE

In order to promote good relations for llamas with other pack stock users, back packers, hikers and the general public, certain rules on packing etiquette should be followed. This is especially true on the public lands of the west where other pack animals, primarily horses and mules, vastly outnumber llamas. These other users of public land can have significant impact on rules and regulations pertaining to the use of llamas on these lands.

Many horses and mules are easily frightened by anything new. Thus, they are often frightened by llamas, particularly the first few times they see them. You as a llama packer need to be especially watchful for horses and mules. Frightened horses and mules can panic and hurt themselves or their human companions.

We recommend that when you see horses or mules on the trail, try to get off the trail and let them have the right of way. If you are close to the riders when you see them, let the riders know that you have llamas and that they may scare their horses. By using these few precautions, we have had no problems with horses while on the trail. Llamas should not be staked-out beside a busy trail that is likely to be used by horses.

If you have dogs with your llamas, be sure that you have adequate control of your dogs at all times. If your dogs charge toward the horses, the horses may react to the combination of dogs and llamas with the llamas receiving the blame.

Llamas are known for their minimal environmental impact; however, you can negate this by the way you camp. You should try to complement your llamas by minimizing your and your guests' effect on the environment. Usually people who pack into the forest and mountains effect the environment most by the way they camp. Any person, with or without a llama, should carry out all trash that cannot be burned. Try to have your camp fires in previously used fire sites, and large, roaring fires should be avoided. Usually there is sufficient downed and dead wood for your fires so standing trees should not be bothered. In some areas, especially in areas of high use, fire pans might be used. In dry summers, forest fires started from careless camp fires or by leaving a camp fire burning is a real concern, and one you should avoid.

You should avoid camping in fragile environments if possible. Do not wash dishes in streams and lakes nor dispose of your wash or dish water in lakes or streams. Human waste, especially anywhere near a camping area, should be buried.

Depending upon the trail, you may want to distribute llama pellets when the animal defecates on the trail. Of course if the animal is moving when he defecates, there is usually no need to distribute the pellets. However, if deposited in a pile on a heavily used trail, you may need to kick the pellet pile a time or two to move the major amount off the trail. In a remote or little used trail, it may not be necessary. When camping, move your llamas often to prevent pellet accumulation in a small area. Pellets should be distributed when llamas are kept in one area.

Llamas are great for minimizing environmental impact. Their soft pads do not damage the trail, and their grazing impacts are much less than for other pack stock.

When llamas are allowed to stand in water, they usually have the urge to defecate and urinate in the water. Many will begin after only a few moments in the water. It is easily prevented by not allowing your llamas to stand in the water. Some llama owners take a can of llama pellets with them, and entice their llamas to defecate before entering the water by letting the llamas smell the contents.

Don't allow llamas to overgraze when you stake them out. Although llamas generally do not overgraze an area, move them often. Trail erosion, overgrazing near lakes or in other fragile or heavily used environments by tied pack stock and horse biscuits on major trails, and rope damage to trees when

TRAIL AND CAMP ETIQUETTE 82

pack stock have been tied are all things that have occurred on public lands. These negative impacts are becoming important issues and attention to these problems will increase as use of our national forests increases. In the near future, expect to see attempts to limit these impacts by restriction of pack stock on certain trails, near streams and lakes. Fortunately, these problems are usually unheard of with llamas. Llamas are well known for being gentle on the environment, and we need to help maintain this positive image.

When meeting other llamas on the trail, try to maintain some area between the two groups of llamas. All llamas and especially studs like to investigate new llamas by smelling them. Invariably the ones being smelled object to the procedure.

Llama owners need to be sensitive to current and future needs of others when camping in established camp grounds or near lakes and other areas with limited suitable camping places. Don't take up all the available spaces with your llamas. You may need to stake them a short distance from your camp. Llama pellets in the center of an established camping spot will not foster admiration for your animals by a future camper. You should also not tie your llamas within 100 to 200 yards of water to minimize the chance of polluting the water.

If you must have your music along with you, use individual head sets. Loud noises have no place on the trail or in camp.

I have been told by forest rangers who have packed with llamas that they do not see grizzlies when they pack with llamas in grizzly country. They thought that perhaps the llamas kept the grizzlies away. This idea may or may not be valid. However, no matter what you are using as pack animals, you should follow all of the rules and suggestions

LLAMAS ARE THE ULTIMATE

about packing in grizzly country established by the park service and national forests. Copies of their rules can be obtained from these agencies.

By being courteous, following these few suggested actions and pausing a few moments to answer questions posed by back packers, other livestock packers and others, you can help promote the use of llamas.

CHAPTER 14

PICKING A PACK LLAMA

Many people have distinct impressions of what the ideal packing llama should look like. Many llamas may not look like our ideal packing llama but are good to excellent packers. For centuries the main purpose of llamas in South America was for packing with wool and meat secondary purposes. So in South America, at least, the llamas were beast of burden. With this rich history, perhaps we should be able to assume that most llamas are capable of doing some packing. However, when you hear a llama producer say that any male llama will make a good pack animal, don't you believe it! Some llamas just aren't suited for packing and will not pack adequately.

As a general rule, the larger the llama, the more he can carry. Although this is generally true, there are some major exceptions. The large, bulky animal, that might remind you of an over-weight lineman in football, generally does not pack very well. So a llama needs height as well as weight. In general, the basketball player-shaped llama is going to do a better job of packing than the football-lineman looking llama. An average llama up to and including the large, tall llama is probably your best bet for making a good packer.

The tall, big, medium bulk animals make tremendous packers. In the west where hunters haul out large mule deer, elk and moose and pack long distances in wilderness areas, it is important that llamas be able to carry heavy loads.

Okay, so the bigger animals make the best packers. What about the small to medium-size llama? There are animals in

LLAMAS ARE THE ULTIMATE

This large packing llama is capable of carrying up to 140 pounds.

this weight class that do make wonderful pack animals if loaded reasonably. For purposes of this discussion, small is classified as under 310 pounds, medium up to 360 pounds. So am I saying that the smaller animals cannot be good packers? Of course not, although in general they cannot pack as much as the larger animals described above. Many people do not have a need to pack the bigger loads that the larger animals can carry. However, the short, wooly, small llamas that are currently popular in the United States as investments are not even going to make the consideration list for good packing llamas. Many, and perhaps the majority of llama breeders, are turning toward the wooly animals in their breeding program, many of which are too small or otherwise not suited for packing. This is not meant to be critical. We also have some breeding animals that produce animals that are large but are too wooly for packing without shearing.

We have some friends that have llamas that weigh at the lower end of our arbitrarily defined medium weight animals. Their pack animals are marvelous pack animals who are kept in shape by frequent use in the mountains or in the hills near their home. Most of their stock are medium to slender built, and wow can they pack! Once Cherrie and I left the trail head in early mid-morning while it was cooler and hiked into a neat mountain lake in two and a half hours. We were not in good shape at the time, but we plodded along with a good effort. Our friends made the trip with their llamas in a little over an hour.

Conformation is important for packing llamas. Since the saddle and panniers ride toward the front portion of the back, and since the rear height of a normal llama is slightly higher than the shoulder height, it is safe to assume that more that half of the weight is carried by the front legs so relatively straight front legs are important. You should view them from

These baby llamas have good packing potential. They have straight legs and backs and have large dams and sires that have produced other outstanding packers.

both the side and front. An animal with knock knees (i.e. front legs that bend inward at the knee) will likely not pack heavy loads long distances as well as one with relatively straight legs. Knocked knees in a young llama will often straighten by the time they are two-years old. Similarly the rear legs should not be excessively cow-hocked (bending toward each other at the knee). Although moderately cow-hocked animals usually make good pack animals. Avoid animals as potential packers that are sickle hocked (rear legs that place the feet well under the body of the animal). Very thin legs are also not desirable in packing stock. The legs must be capable of packing a llama and its load over rough terrain. Strong, well developed leg muscles including thighs are important for packing. Make sure the feet and ankle development is sound and normal. A llama down in the pastern (the portion of the foot immediately above the pad) will likely not be able to pack heavy loads for many years. A relatively straight back is also important. Ear shape or size aren't particularly important to the packing male.

Moderate to short-wooled animals are best for packing, but long-wooled animals can be sheared. The longer wool makes saddling more difficult and also can overheat the animal. Leg wool readily acquires seeds, burrs, mud and other debris and is not desirable.

Buying pack animals is not without risk. Not all animals that look like they would make great packers turn out to be the best, or even acceptable packers. I once entertained a couple of fellows who were shopping for packing llamas. They were impressed with the size of the llamas that we were offering for sale. However, they were looking for cheaper prices similar to the ones they had previously paid. In the course of the conversation, they asked about a llama that they had bought that continually kicked at people. Later they asked if we knew anything about a llama that had a lung or breathing problem and just didn't have the stamina to pack. We hadn't encountered either one of these conditions, but they had purchased both. So make sure you haven't purchased someone else's problem animal. Usually it's better to pay more and get good stock. It is not common to purchase a good-looking pack animal that has physical problems that prevent it from packing. None-the-less there are llamas that cannot pack because of physical problems.

A few years ago, a couple of clients asked me to find them large packing adults. So I called a number of breeders and responded to several ads. Two of the three animals that were found were difficult to handle because of a bad influence on their social development by people who either handled them too early or too often at a early age. Emotional problems that llamas have can and do prevent them from making good packers. A friend of mine purchased a llama that was partially raised on a bottle. He was hoping to use him as a pack animal. It never worked out. Males that are over handled before weaning often just don't do the job as a packer. Others that were mildly over handled may pack but may continually give you problems.

Be suspicious of any packer prospect that approaches you and pushes you. No, this doesn't automatically make him a reject. Some llamas will approach and smell you or let you pet them without being pushy. However, if you can readily approach a prospect or he approaches and touches you, ask questions!

We have a male that is an instant favorite of most everyone that visits our llama farm. He really loves to smell your face, be petted and will walk along beside you. He is a wonderful packer; he gladly packs any position in our pack string; will go anywhere we will lead him including walking up a steep plank into a rocking jet boat in the Salmon River without the least bit of hesitation. If available, he's always on the selection list when we choose llamas for a trek. I think he sells lots of llamas for us, and several breeders have tried to purchase him. So an approachable llama or one that approaches you may not be bad. Many of our packers will let you approach them and if you strike one on the rear, he may not move much. These animals are well adjusted emotionally and fun to be around.

Personality is an important trait in pack animals. However, an out-going, friendly personality may not be necessary. We have a pack animal that would just as soon you not even touch him, but he is easy to handle and is a wonderful packer.

How much should your pack llama cost? I hesitate to answer this question as prices are subject to change. At current prices, expect to pay between $750 to $2500 for a male llama with packing potential. Age, size, training and color all enter into the price. Currently few animals that have potential as large packers are available in the lowest price range.

If the male you are interested in purchasing has been partially or completely trained, ask that he be caught. Then you should at least be able to handle and lead him before purchasing.

I've heard of llamas packing in a commercial pack string until they were 27 years of age. The oldest llama in my pack string is only 11-years old. Therefore, I don't have enough personal experience on the subject; however, I would think that most llamas could pack until they are 20-years old. So don't be afraid to purchase an animal that is several years old. I have purchased animals that were two to seven-years old that had little or no training. One of the advantages is that you often know that he has not been mishandled or mistrained. He has no bad habits as the result of being improperly trained. In addition, the adult animals are very smart and catch on to new things very quickly.

If you are purchasing an animal that has supposedly been trained and is ready to go, I'd suggest some degree of skepticism would be proper. I have heard of too many people purchasing a trained animal only to find out that he will lead a little and that is the extent of his training. If your paying extra for a well-trained llama, ask if he will readily load in a vehicle, has he actually packed out on the trail, how many times has he had a saddle and panniers on, has he been staked out on a cord, has he walked in water, crossed a bridge, etc. Take the llama for a short walk and/or trot. If properly trained, he should move easily with you. Watch the llama move. Natural long-legged, well proportioned llamas have a grace and light movement that is wonderful to watch. Jumping a small obstacle often reveals this easy movement. He should walk and trot with ease. These are all traits that indicate a good packing prospect, and the potential buyer should look for these qualities in any animal being consider for packing.

If possible when purchasing an adult for packing, have him haltered and walk up to him. It helps to determine the size of an individual llama. Also if weight is an important factor ask if the owner has weighed him. When you're purchasing a llama and size and weight are important to you, don't take the estimate of a breeder as to his weight. It is common for many people to state weights of llamas for sale even if they don't own a scales. The weights of llamas are difficult to estimate, partly because of the wool. We've weighed lots of llamas, and we are often surprised at how much our guesstimates differ from their actual weights. We like to play a game with some of our llama owning friends and ask them which side by side llama is the heaviest. Often their guess is incorrect. So if you ask for the weight of an animal, make sure the person has actually weighed it on a scales.

Jay Rais, a well-known llama packer from Boise, Idaho, advises potential buyers of pack llamas to buy from llama breeders who pack. The idea is that usually llama breeders who pack have a better idea of the type llamas that make good packers. We often go llama trekking a time or two with people who live in our area that purchase pack llamas from us. By going once with us, they feel very comfortable in going alone. We also teach llama packing community education courses, and these types of courses are useful to many. Many others do fine without going with anyone or taking courses. Several have done great by following the guidelines in some early drafts of this book.

Color is not overly important, except as a personal choice. Darker animals do get hotter faster, particularly if they have longer wool. Llamas with some white are generally in demand by hunters and, therefore, cost a little more.

The toe nails should be trimmed properly and the animal should have received inoculations and have been wormed, or examined for worms, at least once during the previous year. Also ask what medical problems, if any, the animal has had.

Remember llamas are social animals. An individual llama normally does not do well alone. Although a companion goat or horse might perform this function, two or more llamas penned together or close by is best. If a person can afford only one llama, I often suggest that they go together with a friend and each purchase a llama and then house them together. Also by using each other's llamas for packing, the llamas will have more work and be in better shape and health.

Also packing with more than one llama is more fun and useful.

Purchasing llamas is an exciting part of llama ownership. With a little information, the average person should be able to make an intelligent and wise choice about a packing prospect.

CHAPTER 15

TRANSPORTING LLAMAS

The commercial stock trailer is the most common method of transporting llamas. The so called two-horse trailer can usually accommodate three or four llamas providing you purchase one without or remove the center divider. (Since I object to anyone calling my llama trailer a four-horse trailer, I will hereafter refer to the two general sizes as short or three/four llama trailer and long or multi-llama trailer.) If you desire to purchase a used trailer, this size is commonly available. A trailer with two axles will be easier to pull and hitch to your truck or van. It will also be safer when a tire fails. There are many manufacturers of stock trailers and many models covering a wide range of prices.

Although multi-llama trailers may be harder to locate on the used market, several different lengths and many styles of the multi-llama trailer are available. These larger trailers can haul more than twice the number of llamas.

Stock trailers are made with two general types of hitches, a regular bumper hitch and the fifth-wheel hitch, which mounts in the center of the truck bed. With a regular bumper hitch,

Trailers are a good way to transport llamas. Our 17-foot llama trailer will carry eight or nine large packing males.

you can use the bed of your truck to haul additional equipment and even a stock rack or camper top. In some places we pack, a trailer can't be pulled into an area that can be reached by the four-wheel drive pickup. Therefore, a stock rack or llama topper on the truck permits us to haul our llamas in further than we can get with the trailer. It might take two trips with our truck, if several llamas are being hauled in the trailer. Most people who have used both types of hitches prefer the fifth-wheel hitch as the trailer is easier to attach to the truck and is easier to maneuver. In addition, a truck can pull a much larger fifth wheel than a regular trailer.

If you purchase a bumper hitch trailer, a load leveler hitch installed on your truck supposedly reduces tendencies of the trailer to swing from side to side and does a better job of distributing the weight of the tongue on the truck.

TRANSPORTING LLAMAS 96

In our part of the country, the weather is generally severe during the late fall and winter and very warm in the summer. Therefore, it is important that we have plenty of ventilation in the summer but be able to completely shut off the sides of the trailer in winter. Our trailer has two eight-inch wide openings running along the length of the trailer. In the summer the openings provide plenty of air. During the winter, we place an eight-inch board over each of the openings; thus we make the trailer wind proof and warmer for late fall and winter use. Other trailers are equipped with sliding windows, vents and insulation. You'll need to decide on the type of trailer you need by evaluating the cost, your seasons of use and your particular weather conditions.

Several times, when we have arrived late in the night, we have unloaded the llamas and laid out a sheet of plastic on the soft straw, rolled out our sleeping bags and spent a comfortable night sleeping in the trailer out of the wind.

If you're making a trailer or purchasing one, make sure you purchase one that has a covered top, particularly if you plan to use it in hot or cold weather or plan to haul untrained llamas.

Most of the trailers have maximum inside width dimensions of five and a half feet; however, a few have six foot widths. Each of which works great for llamas. An extra six inches dramatically increases the floor area of the trailer. Thus, the trailer can haul more llamas comfortably with more room to lie down.

When we pack with friends with a trailer load of llamas or for extended packing trips, we need extra storage capacity over that provided by the tack compartment. Since the hay manger wasn't needed for hay, we had the factory completely enclose the sides of the hay manger and to construct a door across the hay manger. This door can be lifted and removed from the compartment. Thus we more than doubled our tack storage capacity. Another modification at the factory was to weld loops inside along both sides of the trailer which provide convenient tie locations.

A tall dividing, center gate prevents studs from unduly bothering females or young llamas in the other compartment. Llamas can be loaded by the smaller side door or the center gate can be swung over to the side of the trailer for loading the front compartment. When the center gate is latched to the side of the trailer, it provides additional space.

Trailers should have several places to tie animals around the outside of the trailer. These various tie locations are handy for spacing the llamas during the saddling operation or for other purposes.

Another factory modification we requested was to modify both doors so that a long-shank lock could be used to lock the animals into the trailer. When we are delivering llamas to a buyer, we wanted to be able to sleep in a motel without having to worry about someone releasing our llamas. Surprisingly, when we were shopping for trailers, many had no convenient way to lock the doors. We keep the locks off the trailer when we are traveling so in an emergency we can remove the llamas in a hurry.

TRANSPORTING LLAMAS 98

The main disadvantage of most stock trailers in hauling pack llamas is their low clearance. For other types of stock, I am sure that this is convenient. Nor is it a problem when transporting llamas between ranches or to shows. However, the trailers' low clearance prevent them from being used on many mountain roads. The longer the trailer, the greater the tendency for the rear of the trailer to drag. Llamas have no difficulty loading into trailers that are much higher. Although I looked for trailers with more clearance, this seems to be an industry standard. However, many of the trailers can be raised four to six inches by companies that service trailers by mounting the springs on top of the axle rather than under the axle. We raised our 17-foot trailer and have been very pleased with the increased clearance, and it did not affect the towing.

By the time several llamas and all the necessary equipment are loaded into the longer llama trailers, there is considerable amount of weight to pull, especially if the trailer is a heavy duty one. The medium size engines in the 350-cubic inch category are marginal for pulling this much weight in steeper mountainous terrain.

Anyone purchasing a trailer for the first time may have difficulty in backing the trailer into selected locations. It is easier to learn if you have clear vision through the back window at the entire front of the trailer. Otherwise it is difficult for the beginner to judge the angle of the trailer by looking only through the right side mirror when the trailer is being backed to the right of the driver. An empty shopping center parking lot is a good place to practice backing a trailer.

The correct operation of electric brakes on trailers is important anywhere but particularly in the intermountain west. Even when borrowing a trailer for temporary use, one should make sure that the trailer's breaking system is correctly wired to the towing vehicle.

Llamas can be carried in almost any type of trailer, and home made or shop-constructed trailers can be suitable for transporting llamas. Llamas do not weigh as much as other types of livestock, and they are tranquil in a trailer so a lighter-weight trailer will do the job and be easier to pull. The trailer should have good tires, pull easily, track the tow vehicle safely and have adequate brakes and lights.

Llamas can be hauled in a variety of vehicles. Llamas can easily jump into large four-wheel drive pickups. A pickup will haul three large males.

Stock racks for pickups are an inexpensive way to transport llamas. Generally three large llamas can be hauled

TRANSPORTING LLAMAS 100

conveniently in a full-size pickup bed. This allows room for the llamas to lie down. For shorter distances, four large adult llamas can be hauled. Two llamas can be carried in a mini pickup. Adult packing llamas can easily hop into the back of a pickup. Used stock racks can often be purchased between $100 and $250 and new ones begin at about $250. One advantage of the normal stock rack is that it can be loaded and unloaded from the truck comparatively quick and easy.

An inexpensive canvas can be custom made by your local canvas maker to go on a pickup stock rack. It gives llamas protection from the wind. Wind, especially on longer trips, can irritate the eyes of llamas carried in stock racks and some type of protection, such as a canvas, is recommended. We had one made that wrapped all the way around the rack including a foot in the back on each side. As I remember, it only cost $30 and had gourmets at the top, bottom and throughout the middle of the canvas for tying to the rack.

Many llamas are transported in camper tops on pickups. In some tops, the llamas have to crawl into the bed of the truck. Then the owners stop occasionally to exercise the llamas and let them "use the facilities" along the road. I do not recommend this particular set up, particularly on rough mountain roads as llamas need to stand up occasionally.

We had a camper top specially made for llamas. It has adequate height for even our taller llamas to stand. It has a regular trailer-type rear door and closing hardware and plenty of windows for summer ventilation. It has extra insulation in the roof to help reduce the heat in the summer. Two large vents in the top also help regulate the heat. It has four inside and four outside metal loops built into the corners to tie the llamas, a large storage area above the cab and plenty of inside and outside lights. It cost about $1400.

Vans are good ways to transport llamas, particularly the larger vans. Llamas, especially medium to smaller llamas, can stand in them. We have had females delivered to our farm for breeding in the back of minivans.

A word of caution is necessary when you tie llamas in any sort of transporting trailer. If the llamas are inexperienced and you must tie them for some reason, be careful and check your llama(s) often. I have heard of several new llama owners purchasing a llama and tying him close to the wall similar to the way you might tie some other livestock, only to arrive with a dead llama. Llamas must be tied with enough lead to allow them to lie down comfortably.

Excited, young untrained llamas have been know to jump out of the opening above the rear door of a trailer. It doesn't happen often, but if you are carrying llamas who have never been hauled, or trained, or perhaps never having been handled, you should be aware of this possibility. Similarly, a wire net over the top if a stock rack may be a good idea if you are transporting new llamas.

Llamas usually lie down quickly when you start down the road. This lowers the center of gravity, and in a pickup stock rack or camper top makes the ride much smoother. The point is that almost any type of pickup truck, van or trailer can be easily adapted to haul llamas.

CHAPTER 16

COMMERCIAL PACKING

Recently we received a request to answer several questions relative to establishing a llama packing business, that is commercial packing involving day, picnic, overnight or several days packing with clients. The information may also be useful to the person planning an outing with friends and llamas.

In the United States, if you plan to pack commercially on public lands, you will need a special use permit from the Forest Service, Bureau of Land Management, or U.S. Park Service. Don't expect to easily find an opening for a hunting or fishing operation. In most areas, a limited number of outfitters are permitted, and usually all the fishing and hunting permits have been taken by other outfitters. Often the only way to obtain a permit for fishing and hunting operations is to purchase an existing permit. Actually, you aren't really purchasing the permit, but only the business, list of clients, and other items associated with the permit. Once you have purchased the business, usually the federal and state agencies involved will assign the permit over to you.

Often in the more popular areas, other types of operations such as photography or nature tours may also be limited. Even if a packing service with llamas doesn't exist in the area, agencies may still not allow any new outfitters in some areas because of full use of the area by other outfitters. However, you usually can find a suitable area for your operation if you are willing to settle for another area.

Federal agencies may require a financial statement from you to ensure that you have adequate finances to begin the operation you propose. They will also require an operations plan. Depending upon the state, you may have to secure a permit and license from a state Game & Fish Department or Outfitters and Guides Board or some other agency that handles this type of permit in the state in which you want to outfit.

The federal and state agencies will set certain requirements for trekking in each area. In some areas, you may be required to provide a fire pan for fires, and in some places, where trees are not plentiful, fires may be prohibited. To find out other requirements such as where you can camp, how far from water livestock can be placed, etc. consult the federal agency in charge of the land where you want to commercially trek.

You may be looking at over one year to complete the application and approval process. On approval you may need to purchase a performance bond to be certified to the state. You will be required to purchase liability insurance before you can expect to get approval for your permit. Expect to pay a minimal annual premium of between $850 and $1,200 for insurance.

Some people rent llamas to people and do not provide guide service. This promotes the llamas and often results in sales of llamas for the breeder or broker. However, if you rent llamas that are poorly trained or that do not have adequate trail experience, you may harm the llama industry. The reputation of llamas as outstanding pack animals will also be harmed. It is likely that many of those who rent such llamas will not purchase a llama in the future. For rental service operations, you'll probably want your clients to take some type of training course. If could be a formal or informal course, and some charge for the courses. Most likely, you'll also need some portable stock racks for pickups in order for the clients to transport the llamas. Charges per llama vary, but are

COMMERCIAL PACKING 104

Although much planning and work go into operating a commercial packing company, it can be fun. Usually the llama treks take place in beautiful country.

generally in the $20 to $40 (US dollars) per day rate with other rates often used for week or monthly rentals.

If you choose to rent llamas to clients or provide picnic or other services on private land without the benefit of liability insurance, then a word of caution is necessary. Please check with a competent lawyer who is familiar with the liability laws of your state or province before beginning this type of service. Even if you have every person who rents a llama from you sign a statement that they have received training in packing with llamas and acknowledging that they fully recognize that there is some inherent danger in packing with animals, you may still be liable if you are sued because of an injury.

Signing a waiver statement foregoing any claims in other situations and industries has not been effective in preventing liability assessments in the courts. The attorney I talked to indicated, "As a general rule, a waiver is not all that effective in preventing successful liability claims. If a person relies on a waver, he is taking a risk with his personal financial future." There are, of course, risks associated with most everything. You should seek competent advice and then carefully consider the risks involved before renting llamas or providing services on private land without purchasing liability insurance.

You will also need first-aid training and certification to obtain a permit for outfitting. Even if it were not required, you would certainly want to have first-aid training. In addition, if you used guides other than yourself, they also will need to be first-aid-trained and some states will require them to have knowledge of the area where you plan to trek before they can be issued a license. There may well be other requirements for both the outfitter and guide so be sure to check with your state or province.

There will be charges associated with filing the application for special use with the federal agency. A yearly fee may also be charged each year you renew the permit. Some states also require a yearly fee for their outfitter and guide licenses. You will also be assessed either a percentage of the fees you charge your clients or a set amount depending upon the land managing agency involved.

Of course, if you are packing with friends or others who are just sharing costs of the trip, then you will not need permission to pack on most public lands. Similarly, you will not need any type of permit from the Forest Service if you are conducting a commercial operation on private land. However, you should check with state agencies to see if you or your guides need permits on private land.

Usually the federal agency will have some rules for the number of clients that you can take out at one time. Even if they don't, you'll need to carefully consider how many clients you can take care of in the worst of weather, personalities, and other conditions, such as age and experience of the clients and how much of each meal has been prepared at home and how much has to be cooked from scratch. Usually one guide per four to eight clients is recommended. However, for safety reasons, many outfitters have a minimum of two guides on each trek. I think many more than eight people would compromise the reasons for wanting to take a llama trek.

For a three to four-day outing, one llama should be able to carry all the personal belongings of two people. You will, of course, have to provide each potential client with a list of all items he or she should bring. This list should be detailed and include suggestions for foot wear, clothes, rain gear and personal items. You'll need to establish a weight limit for these items as well. You may also want to list items not to bring, such as firearms and alcoholic beverages. Many outfitters may bring along alcoholic drinks for after dinner or during the dinners but they prefer to limit the amount of consumption thus avoiding problems. You might also want to put a weight limit on the tent and sleeping bags or furnish your guests with these items as part of the fee or as an extra rental fee. The panniers of one llama with all the personal belongings of two people, their sleeping bags and one tent should usually weigh no more than about 80 pounds.

However, many people want to lead their own llamas and expect that each member of the party will get to lead a llama during the trip. Indeed the llamas are a major reason they have chosen you as an outfitter. So you may want to provide each person with a llama. I suspect that most parties of four to eight people will need two to four llamas to carry the items necessary for gourmet foods and all the necessary cooking equipment, first aid kits and other items necessary for three or four days of trekking. The combination of llamas for personal and cooking gear should result in a llama for each client to lead.

Even if you advertise the treks in terms of mileage covered per day and severity of the trail, you'll want to make sure your clients are aware of the requirements for physical conditioning. Many treks are only moderately strenuous and are done at a leisurely pace permitting many people in their 70's or older to enjoy them.

As far as equipment is concerned, llama halters, packs and saddles are available from a number of suppliers including many of us that are active llama breeders. Expect to pay a minimum of $295 to equip each of your llamas with a halter, saddle and panniers, and stake-out-lead or rope. The remainder of the equipment necessary for cooking and camping will need to be purchased locally or through your favorite camping supply catalogs.

You'll also need a trailer and/or truck to haul a number of llamas to support a llama outfitting business. Planning to transport clients and guides to the trail head also needs to be done. For example, will you expect your clients to meet you at the trail head or will you need to pick them up at the airport or motel?

In addition to cost associated with direct pay for a guide to assist you, remember when planning your operation to include state worker's compensation insurance or some other type of

COMMERCIAL PACKING 108

insurance to pay for injuries to your employees and loss of pay if injured, state and federal unemployment costs and the employer's portion of the social security taxes. These are not options, you have to pay them. Most of these taxes and assessments are calculated on a percentage of the employees' salaries. All of these together mean that you will have to pay a significant percentage of the employee's salary for these costs. Those who are not familiar with running a business may find that paying and keeping records on these items requires a significant amount of time. You may feel more comfortable if you acquire the services of a business and/or tax accountant to help you set up your accounting system.

Trekking with a number of llamas is different than packing with one or two llamas, and the reader is referred to the chapter on pack strings to gain an insight into packing and camping with several llamas.

Charges to clients of llamas treks vary considerably. A partial day's trek for a picnic or day trek might be priced anywhere from $20 to $50 per person. While treks of two to four days usually vary from $55 to $100 per day per client.

Establishing an active, well run guiding service with llamas requires a great deal of time, patience and some money. If you don't have almost full time to run an outfitting business during the months that you desire to operate, you may find that outfitting just won't fit into your schedule. Outfitting is not for everyone. However, if you enjoy talking with and waiting upon others, having the responsibility for the welfare and behavior of clients, enjoy showing beautiful scenes and educating about the out-of-doors and llamas, and you don't mind gambling with some start-up money, then llama outfitting may be just what you want.

LLAMAS ARE THE ULTIMATE

Llama outfitting can be rewarding, both financially and personally. Llama trekking can be an entertaining, enjoyable occupation which takes place in some of the most beautiful country in the world!

CHAPTER 17

SADDLES, PANNIERS AND THEIR CONTENTS

I quickly clipped the lead on the stud and looped the 25-foot tether cord around my hand as I headed toward the tent. I tied him to a tree and repeated the process for the other three llamas. Quickly and without any protest, I cinched their saddles onto their backs. Then it took several minutes for me to empty a couple of the panniers and to double check to make sure my extra knife, sharpener, meat saw, game bags and other items were in one of the panniers. Finally, I put my lunch, coats, water and other personal gear in another pannier. Finally, I hooked the panniers over the cross buck saddles on each of the four llamas, a 20 second per llama task. I was thankful that llama packing equipment is so simple and easy to use.

Quickly gathering up the llamas into a string, I headed downhill to pack the nice bull I had harvested. After a couple hours hike, I approached the downed elk. I tied the llamas to a nearby tree and began processing the elk. After placing the meat in the game bags, I put each meat bag in a plastic bag, dropped each into one of six panniers, weighing each to make sure that the pairs were similar in weight. Each pannier weighed 50 to 65 pounds. My hunting partner arrived about that time, and we each quickly placed a pannier on each side of a llama so that each carried 100 to 130 pounds. We then tied the six by six-point rack onto the top of one of the llamas. We tied a piece of rope across the terminal point of each side of the rack so the ends of the antlers wouldn't drag the ground.

LLAMAS ARE THE ULTIMATE   111

Of course, the llamas showed no fear or alarm at the smell or sight of the meat and stood still as we loaded the packs on them. This is typical behavior of all llamas, even those packing for the second or third time.

Fortunately, putting saddles and panniers on llamas is just this simple. You don't have to be an expert at making diamond hitches and other fancy knots to pack with llamas. Usually the panniers are roomy and hold about anything you need to bring along on your trek. Even if an item won't fit inside a pannier, many panniers come equipped with built-in straps on top of the panniers for tying on bulky items. Each pannier we manufacture and sale even has a strap that attaches to a mate on the other pannier. These are convenient for tying on additional items, particularly those that go across the top of both panniers or for tying a folding table or tent on the top center of the cross buck saddles. I sometimes use the straps to tie my rifle in it's case on top of the lead llama. The end of the case lies along the neck so I can easily unzip and pull out my rifle if needed.

Packing saddles for llamas come in at least three different styles. The cross buck saddle, the decker and the soft pack. The cross buck saddle is usually made of aluminum or steel. The decker is often made of wood. We use the cross buck saddles with good results. Most of the saddles come with two cinches. One cinch normally fits just behind the front legs. Usually the manufacturer suggests that this cinch be extra tight as it does most of the work in holding the saddle in place. The other cinch goes further back on the llama usually four to eight inches in front of the penal sheath.

The saddle you purchase may or may not have the breast strap, the strap that goes from the saddle around in front of the llama. The breast strap keeps the pack from slipping

Cherrie talks to Bandit after saddling him with an aluminum saddle. A breast strap and panniers can be quickly added.

backward. The britchin or rear strap goes from the rear of the saddle across the top and hips of the animal and fits under the tail. The britchin strap keeps the saddle from shifting forward. Llamas are well-shaped to keep the saddle and pannier from sliding too much.

We use only the breast strap with our llamas; however, we do have the britchin strap as an option on the saddles we sell. Many llama packers carry a britchin strap along with them and use it only going down extremely steep places on their trip, if they are carrying heavy loads. At other times, it is either in the pannier pocket or hanging on the rear of the saddle. With

Cicero packs panniers with bags tied on top of each pannier and a roll up table on the crossbuck.

some other saddles, both the breast and britchin straps are an integral part of the saddle.

Purchasing a saddle that can be used without the panniers adds some versatility. Sometimes it is necessary to pack the saddle without panniers. For example, a friend of mine, Mike Tupper, had a contract with a gold mining company to carry five-inch diameter, five-foot long pipe into some roadless areas where they were used in gold exploration. Several pipes were tied directly to the saddle along each side of the llamas. If you need to pack bulky boxes and need to tie them on with rope, you may need to browse through a horse packing manual to determine the best method of tying them to a saddle.

Panniers must be weighable when you have them loaded so look for panniers that are convenient to weigh. Some have handles for carrying them to and from your llamas, and the handles can be used to hang them on the scales. When loaded, each set of panniers should be within one or two pounds of one another. The ones we use have lots of pockets, and we keep quart containers of water and fuel in several of them so it is easy to shift these items between panniers to balance the load. Unbalanced loads make it more difficult for the llama and may make him tire more quickly. In addition, well-balanced loads will not shift on the llama.

Panniers are usually filled before they are placed on the llama. Any tie on items for the tops of the panniers or saddle are added after the panniers have been placed on the llama.

Panniers are made of a variety of materials from canvas to cordura. Some are reinforced at critical points with leather or other materials. Many are constructed of water-repellant material. Some are lined with an additional water-proof material. This double protection guarantees dry food and clothes which is important to me as I sometimes go hunting in the fall for several days away from the road. Often my comfort, if not survival, depends upon our panniers being water proof.

Once you have used panniers with plenty of outside pockets, you'll wonder how you ever got by without them. They are handy for water and fuel containers, sun screen, mosquito repellant, toilet paper, film, maps and any other smaller items that you may need to access while you are on the trail. It really is convenient when you need a drink to reach back, unzip a front pocket and take a quick drink of cool water.

If you plan to trek on rough trails, cross-country or haul out game without the benefit of trails, you may need to shop for panniers that do not hang below the level of the bottom of the llama stomachs. It is important that your panniers do not contact rocks as your llamas walk or drag tree trunks when your llama has to jump them.

Your pannier must be easy to attach to the saddle when loaded with 50 to 70 pounds. If the attachment mechanism is something that has to be operated by your hands while holding a pannier, then good luck! For me, it takes both hands to handle that much weight, and bags that have loops that almost fall in place over the horns of the saddle are ideal. I enjoy seeing llamas packed with simple equipment and few noticeable straps. However, I'm turned off by pictures of llamas with saddles and panniers that have straps going in all directions.

If you haven't previously packed with llamas or back packed, you can purchase some equipment that will make your llama trekking more convenient. If you plan to pack in cold weather, a warm sleeping bag is a must. You can survive lots of adverse conditions if you can get a good nights sleep. In cold and/or wet conditions, a top-notch sleeping bag is a must! Purchase a mummy bag or a modified mummy bag that will stuff into a small bag. If necessary, it should be able to fit easily inside a pannier. Modern filler materials in sleeping bags are light weight and warm. Sleeping pads are also important for a good night's rest. The modern air pads are easy to blow up and are high quality.

A back packing tent is also a necessary item. If you have several llamas and need a large tent for a group of people, then a llama can carry one for you. However, back packing tents are roomy, light-weight and well made. Even ones with room for three or four people often weight less than 10 pounds. You can purchase tents equipped with miniature, collapsible sheep herder stoves. With a few pieces of wood, some kindling and a match, and you can get up to a toasty warm tent in the morning when you are fall or winter camping.

Light weight cooking kits and stoves are convenient for llama packing. Some of the stoves will literally last a life time. We have used our Svea stove for many years. A couple of quarts of fuel will cook many meals on one of these light-weight single burner stoves. Recently, while on a 16-mile weekend trek with Richard Reichle of Dillon, Montana, he used a two burner butane stove that was only two and half inches deep, 12 inches wide and 21 inches long when it was folded. His was hooked to a five-pound tank that furnishes about six days of cooking. Other light-weight stoves are produced that do a great job.

Richard also places two five-gallon rectangular shaped buckets in each of his panniers. His camping supplies are placed in the labeled buckets. Once in camp, he pulls the buckets from the panniers and easily finds what he needs. These buckets also provide convenient seats around the camp fire.

We enjoy having fresh meat, cold drinks, fresh vegetables and salads and cold deserts when we pack in the summer. These are ingredients for gourmet meals. If you are interested in taking some of these conveniences with you, shop for panniers that will hold a large ice chest. Alternatively, you can have straps placed on a metal ice chest and put it on a llama similar to what you would a pannier. However, your saddle must be able to accept this modified pannier/ice chest. Camping with ice and these other conveniences adds considerably to the enjoyment of the trek.

An ice chest is restocked with snow. Ice chests make meats, salads, fresh vegetables and other foods available for cooking gourmet meals.

Ice frozen in plastic containers is an easy way to keep your food cool in the ice chest. Starting the trek with frozen meats, milk and other items will also help extend the effective storage period. If you are conservative about opening the ice chest, you can expect three to six days of cold foods. Cold water poured from the top of the ice in the milk container makes wonderful, refreshing lemonade, kool aid or iced tea.

If you have several llamas, you may want to consider an aluminum kitchen box. A friend of mine made one that attaches to one side of a saddle just as if it were a pannier. It holds all of his kitchen supplies including cooking utensils, plates, flat ware, cups and staples.

One of the most important things that you will want to pack on your llama trek is a human first aid kit. I recommend that everyone, especially if you pack away from the roads, take a first aid class. The red cross or other community agency commonly teach these courses. They can save your life or the life of a loved one or friend. You will also want to include some first aid supplies for your llamas, and these items are covered in another chapter.

A small axe, folding shovel and bucket are items you should also consider. They are often required by federal agencies of their outfitters for fire suppression and prevention. The bucket is handy for packing small items within your pannier. Our wash pan just fits in the bottom of a pannier. In it we place cans of food so we don't have to worry about them poking a llama in the ribs through the pannier.

Normally heavy items go in the bottom of the pannier and lighter items go in the top. Sometimes considerable space can be saved by removing the outside box from prepared or meal-in-a-box type foods and placing the contents in a zip lock bag.

Large items such as tents or roll-up tables, which are very useful and luxurious items, are usually tied on top of the panniers or in the cross of the cross buck saddle. You can also purchase small, light weight chairs that can easily fit on top of the panniers. Sleeping bags can either go in the pannier or on top of them. When you finish packing each pannier, run your hands along the back of them to make sure you don't have any sharp corners that will poke or otherwise make the llama uncomfortable.

Consider purchasing several sizes of rigid plastic containers. They keep breads, cakes, etc. in good shape and they stack conveniently inside the panniers. A smaller plastic container helps keep packets of instant coffee, hot chocolate, tea, kool aid and other instant drinks in one place.

We enjoy cooking steaks, fish, shish kabobs, etc. over an open fire. We pack a grill with folding legs for this purpose. By placing a solid small griddle on the larger grill, we have an excellent place to cook bacon, sausage, eggs and other foods. Dutch ovens are also great to have along and can help provide glorious meals. The aluminum dutch ovens are much lighter and work just as well as the cast iron ones.

If you have several panniers, monographing or using a felt pen to mark some sort of identification symbol on each of them is a good way to keep track of the contents of each. Make a list of items stored in each pannier. Then you can easily find your clothes, lunch supplies, and which bag has the llama food or your dinner in them.

Large zip lock bags can contain items needed for a particular meal. By pulling out a labeled bag, obtaining a few supplies from the ice chest, you can quickly have all the items you need for your dinner.

Robin Clark rinses Kara's hair. A solar shower and a hair wash add considerable to the enjoyment of the wilderness.

Don't forget to take some grain or other treat along for your llamas. We often feed individuals in plastic grocery sacks when we are in camp. Just roll the top down a ways so the feed is showing and set it in front of a llama.

Cigarette lighters make convenient fire starters. Purchase the clear ones so you can see how much butane is left in them. I normally take several along with me so I'll always have a dependable fire source. In wet weather, consider taking along a commercial fire starter or candle to help light the fire.

We have a wonderful miniature, single mantle gas lantern that puts out plenty of light. It comes with a case and fits inside a pannier. Although gas bottle lamps also put out good light, they are more expensive to run and the life of each bottle is limited. Camp candles can also be used, but their light is barely adequate to read by. Flashlights are essential. We use the small halogen flashlights, and they are wonderful for that trip to the rest room at night or for checking on the llamas.

Solar showers are also worth considering. These devices are usually black and when filled with water and laid on rocks in the sunshine, provide delightful warm showers. Nothing feels better than having a warm shower after being on the trail for a couple of days.

Be sure and take a water purification kit with you. Boiling water or using an iodine or chlorine treatment is also an alternative for safe drinking water.

It is easy to forget items that you'll need when llama trekking. The list in Table 2 can be copied and used as a guide when you gather your supplies for a trek.

SADDLES, PANNIERS AND THEIR CONTENTS

A water filtering kit provides good drinking water for the long day's trek.

Table 2. Check list for llama trekking supplies.

Packing and camping equipment

halters_____	plastic bags_____
instant drinks___	water filter kit_____
tether cords_____	flashlights_____
solar shower_____	extra halter_____
rain jacket_____	sun glasses_____
extra lead_____	warm coat_____
contact cleaners_	llama bowls or bags__
sleeping bags____	topo maps_____
grain & feed_____	sleeping pads_____
forest maps_____	curry brush_____
dish pan_____	hand soap_____
First aid kits____	sos pads_____
people___	dish soap_____
llama___	paper towels_____
insect repellents_	sun screen_____
llama___	hat or cap_____
people___	tent_____
tent poles_____	water containers_____
folding table____	ground cloth_____
ice_____	grills_____
toilet paper_____	shovel_____
bucket_____	axe_____
stove_____	fuel_____
lighters_____	fire starters_____
candle_____	cooking kit_____
cups_____	flat ware_____
chairs_____	ice chest_____
lantern_____	

SADDLES, PANNIERS AND THEIR CONTENTS

Fishing equipment
    rods_____
    reels_____
    flies_____
    lures_____
    bait_____

Hunting equipment
    rifle or bow____
    scope cover_____
    ammo or arrows__
    knife_____
    license_____
    meat bags_____
    sharpener_____
    saw_____
    spotting scope__
    binoculars_____

CHAPTER 18

THE LOST LLAMA

If you are a llama packer, you may go many years and may never lose a llama. But if you enjoy exploring remote areas of the wilderness then an old Persian proverb that says "Trust in God, but tie your camel," may be a good one to remember. Loosing a llama can be a traumatic wilderness adventure. Let's examine some causes and solutions.

For whatever the reason, llamas do get loose. However, correspondence with dozens of packers indicate that llamas like to be near other llamas too much to leave their companions and are easily caught again. Some packers take advantage of this herding behavior and let part of their pack string run loose at night, dragging their leads to permit easy capture in the mornings. However, as Jay Rais of Meridian, Idaho discovered, it is necessary to tie the natural leaders in order to keep all of the herd nearby. Luckily, Jay's friend saw the llamas following their leader back down the trail, and Jay was able to get around them and catch the "chief boy" before they had gone too far.

Although not a common event, llamas can run away and llamas do get lost. Llamas can and do get separated from the packer and/or other llamas and either wander off or run off. Most incidents involve either young, inexperienced, new or severely frightened animals.

It's more likely for the entire herd to run off than for a single animal to run away. Mountain lions, which are historic predators of camelids in South America, sometimes cause a string to panic. This happened to Dennis Adams of Ft.

Collins, Colorado and Terry Price, formerly of Phoenix, Arizona while camping in the San Juan Wilderness Area near Silverton, Colorado. They awoke to find that during the night, their three pack llamas had become frightened, apparently by a mountain lion, and all had broken the metal snaps to the screw-in stakes. The llamas, familiar with the trail, had bolted toward the trail head. Luckily, they were found about a mile from camp; their long lead lines were tangled, making recapture easy.

Sometimes individuals become separated from the herd. Bill and Jan Redwood of Mancos, Colorado were camping in Colorado at 10,000 feet one night when a mountain lion came into camp. The entire herd broke free; five llamas pulled their screw-in stakes from the ground and were found lying down about a quarter of a mile away with their stakes and tether lines tangled. Unfortunately, the sixth llama, Pirate, a two-year old on his first pack trip broke the clasp on the tether line when he fled so he had no lead. They spent the next day looking for Pirate with no luck. That night, they could hear the mountain lion screaming near camp and feared the worst.

Jan camped in the area for the next four days, hiking various trails, often with other llamas to help attract Pirate. She finally found the llama's tracks and followed them for two miles. In the mud, she could see where the llama had run and skidded at top speed with the unmistakable tracks of a large lion following the llama. The llama's tracks led over a ledge into another basin, but they were unable to follow them further.

On the fifth day, Jan and her daughter, Anna, were hiking on a rough trail about 10 miles from their original camp. Since there was so much bear and lion sign on the trails, they were carrying a tin can with stones for rattling to warn animals of their approach. Hearing something on the trail amidst the four-foot tall skunk cabbage, Jan thought "O God! Rattle the can." On the rattle as Jan related, "Suddenly a llama stands up and comes running to us like 'Get me out of here!' It was like a man and a woman running to each other in a romantic movie, except the llama was running to us."

Dogs can cause the severe fright necessary to cause even an experienced llama to run away. Two large, vicious dogs attacked the llamas of Judy Fee of Valle Curcis, North Carolina, causing two male llamas to break through fences and escape. One was chased for five miles by the two dogs; this pursuit probably saved some young females. He ran to a store where he was recaptured. The other llama, a particularly nervous animal, just disappeared into the nearby thickly-wooded mountainous area. Six days later, a lady living in the mountains called in response to Judy's radio ad and reported that a llama had been in a meadow near her home for two days. The llama had traveled 10 miles.

Bears don't seem to be a particular problem in spooking llamas. In 1986, while hunting in Montana, my hunting buddy shot a large black bear. When I arrived with my llamas, he already had the hide stretched out on the snow. I took two of my llamas to it. One stepped on the hide and the other one didn't pay any attention to it. Several llama owners reported their llamas, upon seeing bears, were curious but not frightened. In a few cases, a llama gave the alarm call on seeing a bear. (Llamas seldom use the alarm call; it sounds similar to horse's wailing.)

One of Judy Fee's experiences appears typical of llama behavior towards bears. Packing in Tennessee, the llamas spotted a bear in a tree only 40 feet away and watched as the bear climbed down and walked off. "The llamas were totally unperturbed. They acted like visiting royalty, viewing a show put on for their benefit."

Some of my llamas worked all summer in an area with grizzlies. The experienced folks with the llamas thought that the llamas help keep the grizzlies away as they talked to several people who had seen them, and they had seen grizzles in past years. I have heard this from other sources, but I'm still not sure if and how they would keep grizzlies away. However, an aggressive or overly curious bear could be a real problem for even veteran llamas, a fact packers should keep in mind, particularly where bears have become habituated to humans.

The only close encounter Cherrie, myself and our llamas have had with a bear was in Wyoming. We, along with nine llamas encountered a bear about 60 yards parallel to us and near a trail that we eventually needed to travel. The very large bear didn't want to move and moved slightly toward us as we hollered at him. Although they saw the bear, our llamas, even the two-year olds, didn't pay much attention to him.

While other wildlife such as deer, moose and elk may make experienced llamas nervous, young or inexperienced ones may react more strongly. Will Gavin of Bozeman, Montana had a particularly friendly or amorous moose come through camp, causing two of his llamas to break away. He found them the next morning a quarter of a mile down the trail, with one of their long tether cords caught on a rock.

Free-ranging horses and cattle can occasionally cause problems by closely approaching llamas tied near camp. Although llamas usually scare horses that are unfamiliar with them, a few horses may become aggressive. Once while sleeping in the open in the Bighorn National Forest in Wyoming, a herd

This mixed group of experienced and novice llamas pack tranquilly through the flowers after a close encounter with a bear. Bears don't seem to bother llamas.

of free ranging horses stampeded through camp. The noise of breaking limbs all around scared the daylights out of us as we expected to be trampled at any time. Immediately afterward, the flashlight revealed three very calm llamas and one terrified wife!

Although untied llamas usually stay nearby, don't count on it! Even single animals can and do go far from camp. Several llama packers reported llamas going up to 12 miles from camp. Llamas also sometimes decide to leave home. Eric Hoffman of Santa Cruz, California, reported a new llama jumped the fence and walked 10 miles up a trail taken the previous day. Apparently, the llama missed his former residence.

What is the best way to catch a llama that is free? Usually you can just walk up to an experienced llama, but sometimes llamas get confused or frightened and don't want to get caught. In that case, two or more people using several tether leads or cords tied together or a 30 to 50-foot rope will suffice to catch the llama. A bowl or sack of grain will also entice llamas to come to you. Every pack animal should be "hooked" on grain for such an emergency.

Where do you start looking for a lost llama? Most packers recommended searching back along the trail, especially if the llamas are familiar with the trail. If you spot your animals, go around and in front of them. Otherwise, they will generally stay ahead of you all the way back to the trail head.

Binoculars can be a great asset in locating lost llamas, particularly if there are meadows or other openings in the forest and vantage points from which to look. The small roof prism binoculars can easily fit into a small pocket in the panniers.

If your llama doesn't stay near camp, don't give up hope. One year a llama was lost in Montana for several weeks before being found. Other packers reported llamas loose in the mountains for weeks, up to two months. Llamas are capable of surviving on their own, particularly in the spring and summer. Several reported their llamas found lush feed and even gained weight.

Often the problem with a lost animal is that you and the animal just can't find one another because visibility in the mountains is often limited. If compatible with your wilderness experience, bells on llamas could help you find or attract a lost llama. Theoretically, llamas could also be trained to come to a whistle or some type of loud call.

In thinking about the old Persian proverb quoted at the beginning of this chapter, tying your llama so he can't break away will decrease your chances of having a run away llama. Most of the packers contacted do not use any sort of weak link that will break in case of panic and so far have not had any problem. However, I like Eric Hoffman's advice on tying llamas, "Better a llama to catch than a llama to bury." When tied on a 25-foot tether line, llamas can accelerate to full speed before reaching the end of the cord. One of our guides, my wife and I saw our older 430-pound male panic while tied to a 25-foot tether cord on his first trip to the mountains. Almost instantaneously he was at full speed, but luckily the snap broke when he reached the end of the cord.

A friend of mine was not so lucky. His daughter was camping with llamas in Yellowstone National Park in Wyoming. A young llama tied by a tether cord apparently tried to jump a small creek near the end of the cord's radius and instantly broke its neck. I know of two other cases where a weak link would have prevented llama deaths. The possibility of a preventable death of a favorite llama should be reason enough for everyone to create weak links in their tether lines.

Eric Hoffman likes the weak link to be at the end of the cord where it is secured to a screw-in-stake. Then in case of a panic, the cord will break and leave the cord dragging behind the animal. This makes the animal easier to catch. A weak link can also be created by tying the cord to a rubber tubing or weak twine attached to a bush or rock. If you camp where the llamas wrap their tether cords around trees, then the weak link should be near the halter.

Llamas are certainly the best pack animals available for staying calm in nearly all situations, particularly after they have been on the trail a time or two. The joys of packing with llamas, the tranquility they help create in the wilderness and their companionship far outweigh the remote possibility that one of them may run away, and even if one does, you'll probably find it within a few hours.

CHAPTER 19

CATCHING LLAMAS

Whether you have one llama or many, from time to time, you'll need to catch one or several llamas. There are several methods that can be successfully used.

If llamas are in a pasture, the catching may be more difficult than if they are in a small enclosure or corral. However, you can usually train your llamas to move from the pasture into your corral. Even in the summer with green grass up to their stomachs, our llamas enjoy eating alfalfa or baled grass hay. By making some noise and pushing our hay cart into the corral or carrying a couple of grain buckets, the llamas will usually run into the corral. We usually spread out a flake or two of hay so most of them will get a bite or two. Then we shut the gate and catch the ones we want.

In the winter, we feed our llamas daily since we always have a good snow cover. If the llamas aren't already in the corral, they run to the corrals as soon as they see our truck pull in from the highway. So catching llamas in the winter is usually easy.

Sometimes, usually in the spring or early summer, some of the females and their babies will not come to the corral. Therefore, you may need to catch them either in the pasture or by herding them into the corral. By using a long rope between two people, we can usually herd our llamas into the corral. We also use the rope to catch a llama in the pasture. By using the long rope we walk a few llamas into the side of the fence or into a corner. As we approach the corner, we

CATCHING LLAMAS   134

A long rope is used to drive llamas into the corner of a seven-acre pasture.

continually take up the rope to compensate for the shorter distance between the two of us while keeping the rope taunt.

With two people, catching a llama against the fence or corner is quite simple. One person approaches the animal's head and puts his arms around the neck while the other person stands either to the rear of the llama or by the hip and prevents the animal from stepping backward. Usually if the animal has had a little training, they let you halter them easily. With some females that tend to be a little wilder, you can wrap the lead around their neck to help hold them. However, make sure that the lead is wrapped around her neck only once, and you are holding both ends near her neck. Then if the llama gets away from you, you can easily let one end of the lead go and have the lead remain in your hands and not wrapped around the animals's neck. Usually you can keep the female facing you and can manage her until the other person secures her from backing away or until you can slip a halter on her. If the llama gets her rear toward you while struggling to get away, you may have to let her go in order to avoid being pulled about.

The person behind or to the side of her pushes with his hands. I don't think the wool is made for grabbing so I don't grab it, and I recommend that you don't use it as a handle either.

If you need to catch one in the corral, you can use the same method except use a short rope or a couple of leads clipped together. If you are working alone, you can tie one end of the rope to a post and work the llama in between the rope and the corral poles. Many females can be caught just by approaching them or herding them into a corner with your arms or PVC pipe. Usually one person will have no problem catching and haltering one of the females if she has received some training.

Based upon the experience with our animals, it should be extremely rare for a normal llama to spit anytime during the catching operation. We have only one llama that sometimes spits when she is caught. She is also one of the few llamas we have ever owned that would not let us pick up her baby. I am sure that someone over handled her too early as a baby and juvenile. I would be shocked if any of our other females spit on me during catching.

What if you have acquired a particularly wild animal or you want to catch an untrained weanling or baby? Let your other older and calmer llamas help you catch it. By cornering several adults with the weanling or wild llama, their bodies will help restrain the other llama. Then you can reach over or around the others and capture the one you want. This technique works great for untrained weanlings who can go under or through ropes so easily. Catching your geldings or studs in a corral will likely be easier. Most of them are

CATCHING LLAMAS   136

frequently caught and are usually better trained. Usually one person can corner one by holding the lead across your body with the arms spread or just with your arms spread. Once cornered they usually just give in to haltering without moving.

If you have a few llamas or only one in a corral or pasture, you can often use grain to attract the animal to you where it can be caught. In a pen with a lot of llamas, you can be overcome with llamas, with many of them upset because they aren't the ones with their head in the bucket.

If you need to herd animals from one corral to another, a 20 to 25- foot-long flexible plastic (PVC) pipe, usually white in color, can be used to herd the animals. We had two of them which we used. One day when it was extremely cold, I pitched one of them onto the gravel, now we have three of them, one long and two shorter ones.

A PVC pipe can be used to direct or catch llamas in a corral.

Our geldings and young studs graze in a pasture that leads into the area by the barn where there are several smaller corrals with walk-in type gates. Usually 14 to 21 animals are in this pasture. We used a long rope and the plastic pipes several times to push these males into the corrals. Now, the animals often come to the barn if they see us. If they don't, one person can go into the pasture and easily herd them to the barn. Then we open the gates and they file in as if it were a treat to get to go into the corrals. It's humorous to see, and it's pleasing that they were so easily trained to go into the corrals to be captive for awhile.

We do not tackle our llamas if they escape from us while we are trying to capture them. Although you may be able to hold one as you are dragged about and eventually get the halter on, there are some inherent dangers in capturing them this way. The llama or her fetus may get hurt. You may also get hurt. It also excites the animal. It doesn't cause any damage for you to calmly corner her again and catch her correctly.

CHAPTER 20

FEEDING LLAMAS AT HOME AND ON THE TRAIL

The llamas of a casual packer, who uses llamas on the weekend, and/or for a couple of weeks during the summer, will usually receive adequate nutrition and energy at home on pasture grass alone. Most who do not have pasture and need to feed their llamas prefer to feed them a grass or oat hay or a mixture of alfalfa and grass, which has been cut before the stems become thick. The hay must be cut and cured properly to retain its nutrient value. Grass or alfalfa that has dried to a brown color is almost worthless. Since we have both alfalfa and an alfalfa and grass mixture on our farm, we feed grass hay most days and the alfalfa occasionally. We have also fed each on an every other day rotation. If you feed hay that has big stems, your llamas may not eat all of it.

Since the lama species in South America eat forage that is relatively low in protein, many recommend that llamas not be fed high protein alfalfa on a regular basis. Although this seems like a reasonable assumption, research on this is limited, but some is currently being conducted.

I'll outline my ideas on feeding which have worked successfully for many. I'm not necessarily recommending it for your application, but you may want to consider the ideas and, along with other advice, make your own decisions about your feeding program.

Each casually packed adult llama and other non-working llamas should be fed once a day and no more than the llama will clean up in a day. Assuming you are feeding quality hay that is not stemy, they should eat stems and all. If they have hay left over every day or leave the stems, then you are overfeeding them. Llamas, even those of similar size, may have different metabolisms; thus, they may have different food requirements. One of our studs eats considerably less than the rest of the studs while another one eats considerably more. If llamas are pasture fed and have good pasture and have room to move around, they usually will adequately regulate their intake. Remember that llamas are both grazers and browsers so you may want to protect certain shrubs or small trees in their feeding area.

Llamas are much more efficient with their intake of food than other common livestock and pack animals. They are ruminants (they chew their cud) and have a three-compartment stomach. A bale of hay will usually feed eight to ten llamas, males or females, for one day. This, of course, will vary with the size of the bale and the quality of the hay. Usually you can figure feeding 1.8 to 2 per cent of their body weight per day. We normally follow the rule that if some hay is left over, we've fed a little too much, and if every stem is eaten and the llamas act as if they are hungry, we feed a little more the next time.

Avoid throwing hay over the top of your llamas. Hay stems and leaves are difficult to remove from the wool. As previously mentioned in the training chapters, feeding time can be an important training opportunity. By entering the corral to feed adults, juveniles and babies rather than pitching hay over the fence to them, the llamas have an increased opportunity to become better adjusted to people. Since we have good pasture, we feed very few llamas in the summer.

FEEDING LLAMAS 140

However, during the winter, they are on baled hay. During the winter when we feed baled hay, Cherrie and I try to touch each adult and juvenile llama by giving them a pat on the rear. It also gives you a better opportunity to observe each individual's health and conditions of the corral or pasture.

When the temperature gets real cold, that is zero or below at night, you'll probably discover that consumption goes up considerably. They need the extra feed to metabolize in order to keep their bodies at the proper temperature. If the llamas start the winter in good shape, and the females are not nursing babies, and you feed quality hay, they really don't need a grain or other supplement.

If you have pasture grazing that's available in the winter, the llamas will graze, and you'll likely find that the consumption of baled hay is lower. Some owners feed their llamas only a portion of their required needs if some winter grazing is available. This ensures that their llamas are in reasonable condition and not over weight when the packing season starts or when breeding or birth time occurs.

I have never seen llamas paw through the snow on a regular basis to reach the forage as other domestic livestock do. Therefore, once the snow has covered the forage, and is glazed or crusted, I feed my llamas as if they are not on pasture at all.

If you are overfeeding them, and they eat mostly leaves, you may soon have over-weight llamas, and you are definitely wasting feed. In addition you'll have lots of stems lying around to get in the wool.

You certainly need to teach each packing llama to respond to grain, rabbit pellets or a commercially available llama ration or a combination of these feeds so that they will be used to it when it is necessary to feed on the trail. However, if you are feeding your mature llamas, who are not being worked regularly, a regular ration of grain, you are probably not doing your llamas any favors. It's much easier to overfeed them than it is to take the weight off. Perhaps a bite or two of grain per day won't do them any harm. A fat llama has a harder time packing loads in the mountains. The same principal applies to humans when we're out of shape. It's tough walking up into the mountains, but when we are overweight, it is doubly tough and can be dangerous to our health. Similarly, a llama will do a better job of packing with less effort and less health problems if he is not overweight. A consistently fat llama is not a healthy animal!

Because of the wool, you may not find it easy to determine if your llama is overweight. Some weigh their llamas often enough that they can tell when their llamas have gained too much weight. However, most llama owners, particularly if they own only a few llamas, do not own scales. One of the most convenient ways to judge a llama's condition is to examine the breast bone between and in front of the front legs. If the bone is easily seen, your llama may be too thin. If it jumps and trembles like a bowl full of jello when you pat it, your llama may be too heavy. A small amount of padding is usually about right. It's not unusual for a llama to have a little fat deposit during the spring when grazing is prime, and this shouldn't be a worry.

If you look at your llama from the rear, the thigh muscles should be well defined, and you should be able to see some of the stomach as the llama stands. If not, the llama may be too heavy. An obese llama will have deposits of fat just below the tail. Although the pelvis bone and most of the back bone will always feel bony no matter what the condition of your llama, some llama owners feel the ribs and back bone at the withers (above the chest). You should be able to feel bones if your llama is not over weight. By a combination of these, you should be able to determine if your llama is near the proper weight.

Now I'm not advocating that you have a skinny llama. However, a well fed llama doesn't mean having an obese llama. You can love your llamas and treat them with kindness, but overfeeding them should not be a part of that loving care. Llama research veterinarians, among them Dr. LaRue Johnson and Dr. Murray Fowler, have indicated that the greatest clinical problem with North American llamas is that they are too fat! Over weight in the females can bring about additional problems related to successfully getting the animal pregnant and birthing.

In the casual packing situation, the amount of grain you might wish to carry along for your llamas, will depend upon the area into which you are packing. If the grazing is lush and the animals have plenty of opportunity to graze, you may wish to bring only enough grain to give the llamas a treat when they reach camp or to have on hand to assist in a capture if one gets loose. If you're packing into a high mountain lake where grazing forage might be sparse, then you'll need to supply them grain at least once a day. Usually a pint to a pint and a half of grain per animal once or twice per day will be adequate.

Llamas are often used by hunters in the late fall or early winter. Sometimes the grazing is limited to sparse, dry vegetation or the vegetation is covered with snow. In addition, when I am hunting, I don't want to take the time to move my llamas every few hours. If there is any forage at all, I'll move them at least twice per day. So I feed about a pint

Llamas take a feed break on loose leads in the shadows of the Tetons in Teton National Park in the background. A few pauses in lush vegetation will help keep your llamas going.

of grain, each morning and night. Usually I am only hunting a few days at a time and some forage is available. If I were in an area with deep snow or where forage just wasn't available, I'd consider packing in some high quality alfalfa or alfalfa pellets or cubes to supplement the grain. However, try to find smaller cubes as the large ones for cattle and horses will be too big for llamas. If you're feeding only prepared or manufactured feeds make sure the roughage is at least 25% by volume.

While in camp, llamas should be given access to water at least twice a day in the summer or early fall and at least once a day during the late fall or winter. As has been previously pointed out, they may or may not choose to drink the first day or so, but you must offer it to them. If there is snow on the ground, you may find that your llama may be content to eat snow for water.

Some llamas will be packed heavy and worked often during the summer. Some of these are in commercial pack strings or are working steadily packing supplies or used some other way almost daily during the summer. Some operations which I am familiar have llamas on the trail 10 days in a row and then rest them for three or four days. When you are working llamas like this on a commercial basis, I recommend regularly feeding them a supplement, either grain, rabbit pellets, alfalfa pellets mixed with grain, a commercially prepared llama feed or a combination of the above. The grain could be a combination of oats, corn and barley. Most people recommend feeding grains that have been rolled. I'm not sure why this is necessary; perhaps it's just a safe recommendation. I feed the rolled grain for the same reason. For hard-working llamas, you should feed a quart to a quart and a forth in the morning and evening. Rolled grain is light and fluffy and is relatively light compared to whole grains. In addition, grazing as often as possible is recommended. When in camp, the llamas staked out on 20-foot cords should be moved every two or three hours even if it looks like they still have some grazing where they are. You'll find that as soon as they reach a new spot, they'll start grazing. This accomplishes two things; it gives the llama increased grazing opportunity and prevents the llamas from over grazing.

If during the rest periods, the heavily used llamas are kept in a corral, I'd recommend that you feed them baled high-quality alfalfa or an equal quality alfalfa with some grass in it. Feed them all they will eat plus a grain supplement once a day when possible. If they spend their rest time in a good pasture, I'd give them grain once a day. If the pasture isn't the best, I'd consider supplementing them with some alfalfa as well.

If possible, always remove the halter when the animals are in the corral or pasture.

We have had several llamas working this intense schedule, that is three months of packing eight to ten days in a row with a few days rest in between. Some were really worked hard. At least some of them were on the feeding regime exactly like the one mentioned in the preceding two paragraphs. At the end of the summer, the llamas were spectacular looking. Muscles bulged at almost every angle on their bodies. They weighted about the same as when they started the summer, and they were a little plump at that time. That plumpness had been replaced by muscle. These boys were in great condition.

If you're new to the llama world, and would like to know the cost for llama feed, you can estimate it. Hay consumption will be about one and a third to one and a half tons per llama per year if you have no grazing. Cost of hay over the last few years has varied from $35 to $85 per ton in my area. A friend of mine kept records when he had three males, and concluded that it costs more to feed his dog than it did his llamas. As I recall good hay was selling for $40 to $45 per ton that year. $70 to $125 US dollars per llama per year would be a good estimate during an average year, assuming that you have no grazing available.

I am often asked how many llamas can be grazed on a certain amount of acreage. Of course, this varies considerably with pasture quality, growing season, irrigation and pasture management. However, I can relate our experiences in southeastern Idaho. Our frost-free growing season is about 100 to 130 days per year, the pastures were irrigated regularly, but received no commercial fertilizer nor any specific pasture management. Once we placed 25 females and babies in a small pasture, probably two or three acres. The owner of the pasture indicated that the previous year, eight cows had eaten it into the ground, and he had fertilized that year. The grass went to seed on the llamas, they couldn't eat it fast enough. We had about 70 llamas on our farm in 1989 and grazed only about 20 acres of our farm. During the same year, we purchased polypay sheep to help eat the grass and to supplement our income. So in our situation, several llamas can be grazed on each acre.

Good pasture management under an intensive rotation system would permit us to graze many more llamas and sheep. We are introducing such a system this year. We report on pasture management in a later chapter.

Many breeders supplement their young llamas by using some sort of creep feeding arrangement that prevent the adults from getting into the area where the supplement is eaten by the youngsters. During the spring and summer when high quality grazing is available, it probably isn't necessary.

Nursing females don't really need any type of supplement in the spring and summer when it is warm and good grazing or good hay is available. However, females that nurse babies in the winter may need some additional high-energy feeds, especially during colder winters.

If you plan to feed hay sometime during the year, you'll need to think about the type feeder you want to use. Several years ago, we built some feeders from mail-ordered plans. Each feeder had two or four key-shaped holes for the llamas to stick their heads into. The feeders do a good job of preventing waste. However, if you have many llamas, this type of feeder is not practical.

We have seen several feeders that have a v shaped manager for the hay mounted over the top of a split 55-gallon barrel. Similar wooden ones are often created against a wall. Supposedly the leaves and stems fall into these containers when the hay is pulled out of the manager. Most of the ones I've seen in operation fail miserably in this regard and waste lots of hay. If you have the manager mounted too high, when the llamas pull out hay, they will get it into their wool. I'm not familiar with all of the commercial ones, but the ones I've seen haven't worked all that well. If you are interested in a commercial feeder, try to find someone who has some similar to it and visit them to see how it actually works.

Some owners are innovative about the feeders they construct or their modification of other equipment. A grocer I met modified some grocery and meat displays into excellent feeders. As another example, we've had excellent results by using tractor tires. On one side, we cut the side wall off. This side is then placed on the ground so that water will not accumulate in the tire. You might desire to place some plywood under the tire. Then the hay is placed in the tire. Several llamas can feed at each tire, and there is very little wasted hay. Usually the tires can be obtained free at tire shops. Some of our tires are brand new--nothing but the best for our llamas! Actually they were factory defects so were free. In a dry climate like Idaho Falls, the tires work great. You should consult your veterinarian or other animal experts in your area before feeding on the ground.

When we feed in the winter, we often put hay on top of the snow out in the pasture. We feed in a different, clean place each day. This forces the llamas to get some exercise and prevents build up of feces and stems in the corrals. Hay carts mounted on bicycle wheels can be purchased that make feeding a bale or two of hay in the pasture easy.

You will need to furnish your llamas with salt and perhaps other trace minerals as well. Some prominent llama veterinarians recommend that you feed supplements in a grain mixture or hay-type pellet. Others suggest a ground salt-mineral supplement. Apparently, some llamas do not adequately use salt blocks; thus in selenium deficient areas, they may not receive enough selenium. Our llamas use salt blocks very well. We use block salt which contains a number of trace minerals such as selenium, zinc, iron, manganese, copper, iodine and cobalt. Usually salt blocks for a particular area have been formulated for deficiencies for that area. However, you may want to have your feed analyzed to see what elements are lacking or in short supply. You may also want to have a blood sample from some of your llamas analyzed to determine if your llamas have any deficiencies. Then you can have a supplement made specifically for your llamas' needs. Your veterinarian or county agent should be able to advise you where to have both analyzed.

Your llamas must have access to good, clean fresh water. Water troughs can be economical to install. An old bath tub makes a good, inexpensive waterer as do the round and oval, stainless steel or rubber water troughs available at many livestock supply stores. If you live in a cold climate, you may need to purchase a thermostatically controlled heater that floats on the surface of the water. Floating styrofoam over a portion of the tank can help prevent heat loss in the winter and help keep the water cleaner in the summer.

When we purchased our farm, we installed automatic filling frost-free waterers. These consist of a three-foot tall cylinder with a bowl of water on top. The water has a thermostatically controlled heater. A cement pipe below the waterer is a key to it being frost free. This cement pipe extends six feet below the surface creating a space for heat from the ground to rise;

thus, keeping the plastic water supply hose from freezing. They provide year-round water with little effort. Cleaning the bowls a few times per year is all that is necessary. This system usually keeps water free of ice, even in below zero temperatures. However, if you live in an extremely cold environment, a light bulb can be installed inside the cylinder below the ground level. When the weather is extremely cold, you can switch on the bulb for a few days.

With a little planing and care in your feeding program, you can keep your females and packing males in excellent shape without getting them too fat or wasting feed. In addition, feeding time can be the most enjoyable part of your day. It does wonders for me as I quickly forget about the problems associated with my job and other activities. Feeding time can be a great mental health activity.

CHAPTER 21

BARNS, SHEDS AND WINDBREAKS

You should give serious consideration to providing some type of shelter for your llamas. Often you can use an existing structure, as is or modify it slightly to suit your needs. Many new llama owners need to construct facilities. These range from the inexpensive and high utility to the expensive and elaborate, serving individual tastes and budgets.

Before the beginning or prospective llama owner assumes that llamas are fragile creatures, in constant need of shelter, we need to establish something about llama hardiness and their comfort range. As I understand it, llamas in South American primarily inhabit areas that have cold winters, summers with warm days and cool nights and dry climates. Since llamas have developed within this climate regime, it is probably safe to assume that in North America, they are the most comfortable in areas with these similar climates. Since llamas live in every state and province of the United States and Canada, we can further assume that they are adaptable to much wider climate zones, especially if we provide shelters that help facilitate their comfort.

We can further make the case, that even within their normal range in South America, there are many days when shelter would be beneficial. Of course, many llamas exist in hostile environments without the benefit of shelters. Similarly in many severe climates of North America, sheep, cattle, horses and other domestic stock survive outside without the benefit of shelter. Are llamas more fragile or more susceptible to the extremes than those creatures? Probably not (with the exception of heat and humidity, which will be discussed later). So are llamas capable of living outside without shelter in severe cold, wind and snow or rain. You bet! Many llamas used as sheep guards in western states are guarding flocks and survive every bit as well as the sheep, even in 20 to 30 below temperatures.

During severe storms, livestock would also likely benefit from shelter. However, for many livestock operations, it is just not practical or affordable to construct shelters. If they can be fed properly, they'll thrive in winter conditions. So why worry about shelters for llamas? There are several reasons why it is prudent for your llamas to have access to shelter, even though they could survive without it. No matter what type of severe weather that is prevalent in your area, it can stress your llamas. Although I live in a cold winter climate, the cold really isn't as much of a concern as is strong winds when it is extremely cold. So shelters help protect and provide comfort for my animals.

Depending upon climate, some llamas need protection from the cold and snow, others need a retreat from the heat or drying facilities from constant rain. In some climates, llamas really don't need protection.

Protection, at least from cold, is not needed full time. During a previous winter, snow melted in Idaho Falls in January which is unusual. Our high temperatures were mostly in the teens to 30's which is a relatively mild winter and only a few days were below zero. My llamas were not permitted into the shelter a single time. They did have an adequate wind break. Our llamas commonly lie down next to our sheds and barn which provide wind protection even if they aren't open for use. In addition we have over 200 trees near our corrals and the nearby neighbors also have windbreaks that help reduce

BARNS, SHEDS AND WINDBREAKS 152

Barns can provide increased comfort to your llamas in severe weather. They also keep hay and other feeds dry. This one has a connecting shed as well as indoor corrals.

the winter winds. Native forest and shrubs provide the same protection. You may want to consider planting a wind break on your land. If you live in an area that has cold winters, and your llamas are in a pasture without any natural or man-made windbreaks, by all means provide your llamas with winter protection! Llamas will eat needles and bark from many conifers, but they leave the bark of our deciduous trees alone. You'll have to protect young trees from llamas until the foliage is too tall for them to browse. Llamas like the leaves of most deciduous trees and will browse the lower leaves. My llamas love the leaves of cottonwoods, and in the fall seek them out as the leaves fall from the trees.

Most llamas in temperate to cold climates are provided full-time access to shelters. This much protection is really not needed, but it certainly isn't harmful. However, one of the favorite places for llamas to have a dung pile is in shelters. In our type of climate, these piles will be frozen solid from November to early April. So I prefer to have relatively clean sheds for the llamas to use only when necessary. We don't particularly enjoy cleaning out sheds so this is a reason not to let them use the sheds full time. A small chain across the entry keeps most llamas out of shelters when the weather is mild; wider openings may require two chains. It is not cruel to let your llamas be outside in cold weather if it isn't going to snow heavily and if they have shelter from the wind.

Several types of sheds and windbreaks can be used for llamas.

Since the cold, relatively low-humidity winters of the intermountain United States is not too dissimilar from the llamas' range in South America, llamas have no problem with the climate. Also llamas do great in the warm, dry climates. However, the other extreme, hot, humid climates in the summer can be dangerous to your llamas. In these climates, llamas should have access to shade during hot, humid weather.

BARNS, SHEDS AND WINDBREAKS 154

Wet sand, sprinklers and shearing can and are used during days of extreme heat periods and humidity to prevent heat stress.

Llama wool was not designed for continual wetness and problems can develop if the wool is continually exposed to wet conditions. So if you live in a wet climate, such as some of the coastal areas, at least have a roof of some type for the animals to get under.

In our cold climate, a three-sided shelter is usually sufficient to protect llamas. Of course the open side should be toward the east or at least in a direction where the wind is less likely to blow. A dirt floor works fine. A portion of most any type of barn or hay shed for hay and storage can be used by llamas. Our shelters have dirt floors and sit upon and are tied to a few cement blocks, which are mostly buried in the soil, to prevent the shelters from blowing over. These types of shelters are relatively inexpensive. Most libraries will have books that contain building plans for barns and shelters. Also look around your neighborhood for other plans and barns that might suit your needs.

If you plan to have crias born in the winter or purchase animals that may be bred for winter births in severe climates, a small heated nursery may be necessary. We built one that was inexpensive.

Avoiding weather-related stress may add a few years to your llamas's life, will ensure that your animals remains healthy and if you can prevent winter-related stress in your females, it should improve your changes for normal and healthy crias. Of course shelters for new-born crias are essential in some climates. Having adequate shelters during storms may also decrease food consumption.

CHAPTER 22
FENCES

If and when you make the decision to purchase llamas or if you decide to move your current llama operation, fences will be one of your primary concerns. You will be lucky if the place you want to keep your llamas is already fenced with good quality fence that you can use immediately.

Many different types of fences will keep your llamas safely inside your property. A friend of mine, Don Trupp, has raised several different types of domestic livestock. He indicates that llamas are much easier on fences than any of the other stock he has raised.

It really doesn't take much of a fence to keep llamas at home. Although llamas have the ability to jump, they seldom jump over fences. If your llamas are reasonably happy and reasonably accustomed to people, then the fence doesn't have to be the greatest. However, studs next to one another in the presence of females is a major exception. When I purchased Bandit, one of our studs, he was 13-months old and was in a pasture with three other young males, none of which had been caught more than once or twice in their lives. While we were catching Bandit (which was a difficult task), two of the other males jumped a four-foot barbed wire topped fence. The males were severely frightened by people since they had seldom seen anyone or been handled so they jumped the fence. Similarly, I have seen studs jump wooden fences to fight with one another. So there certainly are a few exceptions.

If you need to fence an area for your llamas, several alternatives are available. For most fence options, fence posts are required. Usually in Idaho, you can assume than an untreated post will remain strong and functional for seven to ten years, a creosote-soaked post will last a few years longer. A pressure-treated post may do the job for 30 years or more. Many variables such as winter conditions, rainfall, pasture irrigation, etc. will effect the life of posts. Obviously the pressure-treated posts are the best bet. Building fences is expensive and hard work; something we wanted to do only once so we used pressure-treated post.

Barbed-wire fences are one of the least desirable fences. However, if you purchased a place that has barbed-wire fences, they can be used effectively if you are sure you don't have studs next to one another. Barbed wire pulls out wool and can cut your animals. When we used to rent, we used two pastures partially fenced with barbed wire without significant problems.

Pole pasture fences are attractive but are more expensive. For large pastures, they may not be very feasible unless you live in an area where poles are plentiful. Woven wire fences do a good job. Often a woven wire fence is installed with a couple of single-wire strands above them. This type of fence is common in many areas.

One of the better options is a high-tensile steel wire fence. The single strands can be with or without electricity. A client of ours in Washington state built a fence for newly-purchased animals out of steel wires which had been used to support his grape trellises. His fence was then tightened with built-in fence tighteners operated by a wrench. He has a very functional and inexpensive fence.

More and more llama breeders are turning to an electrified high-tensile fence, often called a New Zealand electric fence. The electric New Zealand fences have several advantages. It completely stops physical interaction across the fence; works well for a variety of animals and can be predator proof if enough strands are used. With high tensile fences, a fence post is required about every 80 feet or so with a couple or three stays in between them. Our stays are round 2" diameter fiberglass posts with holes drilled through them for the high-tensile wires. Since the wire is stretched, substantial corner posts are required. Ours are eight-inch diameter posts in 38" deep holes and of a double H design. The corner posts are also braced with an H brace in both directions and a brace toward another post at right angles to the corner. Several companies specialize in selling supplies for this type of fence. Several of the electric fence companies listed in chapter 33 offer booklets on New Zealand fences and installation instructions. One of the better booklets entitled "The New Fencing Systems Made Simple" is available from Premier Fence Systems.

If you have studs penned next to one another, and females are within sight, there may be more fighting through and over a conventional fence than what you or the fence will tolerate. The fence must be strong enough to withstand the constant charges. Don't under estimate their potential to rise on their hind legs and to weight the fence down with their necks and shoulders. Some put up two fences a few feet apart so the studs can't reach one another or they do not put their studs next to one another.

I highly recommend an electric fence. We have six stud corrals with attached pastures separated by New Zealand electric fences. These five strand high-tensile wire fences feature alternating hot and ground wires just like the five- or

Electric high tensile fence, often referred to as New Zealand fence, is a practical fence for llamas.

seven-wire fences surrounding our pastures. The studs definitely respect the fence, they don't fight over or through it nor do we hear any vocalization that often occurs when studs are highly disturbed with one another. The electric fence makes for quiet, well behaved neighbors.

The seven or more strand New Zealand fence is predator proof for coyote, dogs, and bobcat sized animals. The fox that ranges over our's and the neighbor's farm goes through the fence without too much difficulty. In my opinion, one of the best things about electric fences is that most people are afraid of them. Thus, the fence keeps unwanted or uninvited guests from entering your property at least via any electric fence. We also have electric gates.

Corrals can be constructed from a variety of materials. Although the sides of our stud pens and the attached pastures are New Zealand electric fences, the front of the corrals are made of wooden rails. In our three main female pastures-corrals, the corrals are made entirely of wooden rails. Like the wooden posts on our pasture fences, both the posts and the rails of our corrals are pressure treated.

There are other options besides New Zealand and wooden corrals. In some areas, used pipes or cables are in plentiful supply. For example, some of the southwestern states have areas where oil well drilling, development and pumping are prominent industries. Pipes and cables can usually be bought cheaply. The pipes are then welded to make strong, attractive and permanent fencing. Synthetic materials, although usually more expensive, are becoming more plentiful and diverse.

CHAPTER 23

PASTURE MANAGEMENT

Pasture for your llamas isn't really necessary as many llama owners and breeders do not have pasture and feed their llamas all year on baled hay. However, grazing on pastures is convenient, economical and a healthy way to feed llamas. With proper management, a pasture will provide your llamas with years of grazing. Proper management is the key. Although I am certainly not an expert in pasture management, some basic guide lines can be outlined that may be useful to the novice.

If you need to plant a pasture, contact your local county agent or the local farm seed supply company for advice on what species to plant for your location. Usually a pasture with several species of grasses does better than a pasture with a single species. Llamas can also be grazed on croplands after the harvest is completed. Late in the fall and early winter, llamas can also be grazed on hay fields.

We will assume that you either have an irrigated pasture, if you live in the arid or semi-arid portions of North America, or your area receives adequate rainfall to produce good pastures. A few llama owners successfully use large fenced range land for pastures. Llamas that guard flocks of sheep are commonly found in the range-land type grazing situation. However, the vast majority of llama owners either have no pasture or use planted or seeded pastures.

Many llama owners have only limited pasture available that will not support their llamas for all the grazing season. Often in this situation, owners supplement their llamas while they

PASTURE MANAGEMENT

With proper management, a pasture should produce excellent forage for many years.

are on pasture by feeding their animals baled hay. If green grass is available, this may not be very successful as the llamas will usually eat the leaves and other choice parts of the baled hay and waste a great deal of the rest. Nor will it prevent them from over-grazing. If you don't have enough pasture, it may be better to rotate the llamas between corral and pasture.

In the spring, if you have limited pasture, it will be more productive if you allow the grass to reach several inches in height before you begin to graze it. Similarly, when the llamas have grazed the grass to within three or four inches of the ground, remove the llamas and allow the grass to recover. Just how close you can allow your animals to graze depends upon your location and the species of grass, but for most grasses this is three or four inches. Some of the shorter growing species, such as bermuda grass, should be grazed shorter. The amount of recovery time will depend upon your climate, irrigation or rainfall and the season of the year. In general, grass grows faster in the spring than at other times so the grass is more tolerant to close grazing. In the spring, it might take only three weeks, while later in the summer, five or six weeks might be necessary for the grass to recover. You can usually graze the forage shorter in the spring than later in the year so leave the grass longer when growth is slower.

You may be fortunate to have a larger amount of pasture available for your llamas. In a larger pasture, it's a common occurrence for livestock to continually graze one part of the pasture almost into the ground while allowing another portion of the pasture to become tall and less palatable. Similarly, if you have a pasture of sufficient size, your llamas will primarily graze in an area with the best tasting plants. Most grazing animals prefer lush plants, that is the stage of growth which is the earliest and contains a high leaf to stem ratio. When the plants begin to become stemy and develop blossoms, they begin to loose some of their nutrient value and taste qualities. Plants that are beginning to set seeds are generally the least palatable to your llamas. In a larger pasture you'll find grass and forbs in all three stages of development. The llamas will avoid eating the mature plants if other more lush growth is available. Usually the lush plants are located where they have previously grazed. Often when grass is continually grazed so short that more than 50% of the leaf area is continually removed, the vigor of the plant is drastically reduced. As part of this process, root growth is also drastically reduced. This encourages the invasion of weeds and other less desirable plants. If you have llamas continually overgrazing an area of your pasture, you may need to consider methods to improve your pasture management.

To avoid overgrazing some parts of the pasture and undergrazing the rest, many North American livestock owners practice a system of rotational grazing. This method of controlling grazing forces the stock to more uniformly graze a pasture. New Zealand, in particular, has for many years practiced efficient rotational grazing. In rotational grazing, pastures are divided into a number of smaller pastures. Each subpasture is extensively grazed for a short time. Then the animals are periodically moved to other pastures. Plants are allowed to recover between grazing periods and all the plants are grazed. Thus you obtain increased forage from the same amount of pasture.

The information in this chapter isn't intended to be applicable to range lands of the western United States, particularly those that are in less than good condition. Short duration grazing or rotational grazing systems on range lands has sparked controversy and scientific studies since Savory published his Ph.D. thesis in 1979 on the subject. Many of the studies are impact related, dealing with such subjects as livestock performance, watershed ecology, plant dynamics, wildlife, etc. I took the time recently to review the results of several of these studies reported in the Journal of Range Science. The results on range lands are mixed. However, I do believe that the system has been used successfully on seeded pastures.

Animals in a pasture effectively harvest or reduce forage in several ways. 1) grazing, 2) walking, running, lying and 3) deposition of urine and feces. Fortunately, llamas prefer to deposit their waste in a few locations so feces and urine contamination is not as much a deterrent to foraging as in other livestock. However, you will probably discover that the same llamas that choose only one or two locations for dung piles in the corral, may have a dozen or more of them in a pasture. Walking, running and laying are also harmful to grass and will result in decreased forage production.

Recently I spoke to Pam Thier of Boise, Idaho who is using a more intensive system of rotational grazing called strip grazing to graze 30 sheep on two acres. She explained that they provide a three-foot strip of "new grass" to their sheep each morning. It takes them about 15 to 20 minutes to move their two 300-foot long portable electric fences. The sheep rapidly eat the grass on the strip before it becomes trampled by feet or fouled with feces or urine. The strip is actually six-feet wide, but only three feet is "new grass." This method doubled the length of their grazing season over their previous system of pasture management.

According to Bob Kingsbery (Sheep Magazine), continuous grazing by domestic livestock results in about one-third of the available forage being eaten. Rotational grazing by subdividing a pasture into three or four pastures improves the harvest to 50%; by dividing the pasture with portable fences into areas to be grazed for two-three days each, the harvest potential is increased to 60 - 75% of the forage. By using the strip method with two portable fences and moving the animals every day, animals can potentially harvest up to 90% of the forage. By comparison, mechanically green-chopping may harvest up to 95% of the available forage. These estimates may or may not be correct, and he did not present any study

PASTURE MANAGEMENT 166

Llamas deposit their pellets in common locations in the pasture or corrals. These accumulation areas should be periodically cleaned.

evidence to support the conclusions, and they would likely vary between locations. However, few people could effectively argue that the ranking of the grazing methods was not correct and that substantial improvement in quantity of forage harvested and health of the pasture are not obtained by some type of rotational grazing.

We plan to strip graze some of our pastures with a flock of polypay sheep by installing two electrical portable fences about 6 - 12 feet apart. The sheep will reside inside the two fences, and a portable water tank will supply them with plenty of fresh water. Since we already have New Zealand electrical fences, it will be easy to attach the portable fences to the permanent electric fences. Every few days, we will move both fences three to six feet into the ungrazed grass. Thus about half of the newly occupied strip will be ungrazed and the other half of the strip will be grazed. Under this system, animals eat even the usual undesirable species such as thistle and knapweed. Llamas will eat these plants even under a moderate grazing system. My llamas eat the blossoms of thistle and if you can prevent seed set, some day you'll have thistle-free pastures. The state botanist for the Bureau of Land Management told me about a herd of llamas that had cleaned out some pastures of knapweed.

Am I advocating that you place your llamas in a 6- to 12-foot grazing strip that is to be moved every day? Although I think it is feasible, it is not very practicable. Llamas are larger animals and need room to move about. Although we are going to use the strip method with the sheep, we will use a rotational system for the llamas. Instead of having our females divided about equally into three pastures, we concentrate most of them into one pasture then rotate them through each of the pastures. In the larger pastures, we have used a portable electric fence to keep a few llamas in about one-half of the pasture. The llamas are changed to different pastures as needed to provide proper time for the grass to recover.

You may decide that you also need to improve your pasture management by rotational grazing. To accomplish it, you may need to subdivide your pastures with cross fencing. However, in our situation, I wanted them to have access to a full pasture during the late fall, winter and early spring. In addition, portable electric fencing provides greater flexibility in letting me choose where they are to graze or not to graze. Portable fencing is also handy for protecting a small reseeded area until the plants are well established or to protect a stream bank or

PASTURE MANAGEMENT 168

to force the animals to graze in an area that needs some brush control.

Although New Zealand high tensile electric fences have been used successfully by many llama owners, portable electric fences may or may not be suitable for all llamas. Portable electric fences are more dangerous because it is possible for a llama to get entangled in a portable fence. Llamas do not get tangled in the high tensile fences. Although we have used it successfully in small areas with sheep and a few llamas, we intend to experiment further with portable fencing for llamas. My llamas don't mess with the New Zealand fences; they completely avoid them. However, if you do not have electric fences and want to try the portable fencing, you will want to set up a portable electric fence in the corral and get your llamas used to it under your close supervision. This is extremely important! Your llamas need to be well acquainted with an electric portable fence before you enclose them inside it. You will be better off having permanent fencing to divide your land into smaller pastures.

Although cows and horses can be effectively fenced with one strand of electric wire, we will need three strains of wire for llamas. Portable electric fences need a good ground, and irrigated pasture will usually provide adequate ground. If it does not, a ground wire may need to be installed in the middle of the portable electric fence somewhat similar to that described in an earlier chapter for the New Zealand electric fencing.

Portable electric fences can be moved fairly quickly. Most use some type of polywire or polytape. The polywire consists of a combination of plastic and wire and comes in several colors. The polytape is made of similar material but is wider and more visible. Both can be purchased on a portable reel that quickly attaches directly onto an existing high tension electrical wire on a New Zealand type fence. Some have three wires on the same reel, permitting quick installation and movement. Portable fence posts are made from plastic or fiberglass and have build-in insulator slots for the polywire or polytape. The portable posts are easily installed by pushing them into the soil with your foot. By moving one or two posts at a time, the portable electric fence can be moved several feet without even detaching it from the permanent electric fence.

If a New Zealand electric fence does not exist, a single high tension electric wire can be installed along existing perimeter fences using offset insulator mounts commonly available from dealers who specialize in New Zealand fences. Your portable electric fence can then be "hooked up" to this perimeter wire.

After you have rotated your llamas into new grass a few times, they will eagerly move into a new grazing area with little assistance. You may have some additional water supply duties if you divide your pastures differently than they now exist. You may need to provide them with some temporary water troughs. Llamas must have continual access to fresh water. Filling temporary troughs in the spring and summer takes a little time every few days, but it is easily accomplished with a water hose.

Your pasture management program needs to be flexible to be successful. Weather patterns, such as an unusually warm and early spring, a drought or unusually hot conditions, can effect your pasture growth. Grazing time in each pasture should be adjusted in response to these conditions. If your going on vacation, just open up more pasture to the llamas.

The dung piles of the llamas in the pasture need to be removed occasionally. We normally clean our pastures once or twice per year. The pellets are excellent fertilizer for crops or gardens. Ground or whole pellets are excellent indoor fertilizer for house plants as it has little or no odor.

We have a rental home on our farm that has a huge grass lawn. I am considering using portable electric fences to allow my geldings and/or sheep to graze the lawn occasionally. This beats mowing. Orchards or the grass along the lane on your property or the grass on the ditch or near your barn could also be grazed by the use of portable fencing. However, please use good judgement in where and with what animals, you wish to graze in this manner. You don't want to put your animals in any danger.

Somewhere in a dryer part of your pasture and/or corrals, the llamas will pick a spot or two to use for dust baths. Soon you'll see a depression formed as each day the llamas will roll and flip dust all over their beautiful wool. This is normal behavior and supposedly it helps to keep their wool healthy.

By using good pasture management techniques, you can protect your pasture from overgrazing and undergrazing, force your llamas to graze in a particular area, enhance stream banks, shrub hedges, beautify the pasture near your barn by permitting it to grow, reduce erosion, improve your pasture and extend its life and increase the amount of forage available to your llamas.

CHAPTER 24

WITH LLAMAS, PACKING GAME IS FUN

I turned the spotting scope up to its maximum 45 power to get a better look. Ahhhh- a bull! After hunting eight days without seeing a bull moose, it was gratifying to see the antlers on the moose lying far off in the distance under a fir tree. He was high on the slope across a deep canyon, but a road wound down into the bottom of the canyon.

After installing two sets of chains on the four-wheel drive, I slowly began the crawl into the bottom of the canyon. The bull was located near the top of an old logged over area. A laborious hike in the deep snow brought me into the vicinity of the bull, some three hours after I had originally spotted him. As I approached the area where I thought he would be if he hadn't decided to move, he suddenly ran across in front of me. I fired three times as he covered the 75 yards to the thick trees. Finally, after showing no signs of being hit, he went down. All three bullets had struck vital areas. I was exhausted from my climb up so I sat down on the carcass to rest.

As I rested, I contemplated what for me is now one of the real joys of hunting big game - in this case packing the moose down the mountain in the deep snow. For many hunters, this would be a significant problem and definitely not a joy to haul the moose carcass. Although I was hunting by myself and was several miles from a road, it posed no problem. For back in camp, I had four llamas, all eager to hike after spending so many days in camp. Many consider packing out carcasses a real drag. However, since we acquired llamas, hunting was never so much fun. Now packing out the meat and head with these tranquil and cooperative creatures is a favorite and pleasurable task.

Although none of them had been around a dead moose, I knew that they would pack it without the least bit of hesitation or fear. The week before, a friend from Washington, Larry Cadwell, had shot a hugh black bear near Hebgen Lake. He had skinned the bear and had the hide spread on the snow when I arrived at the scene with two of my llamas, neither of which had seen a bear. I walked over to the bear. One of the llamas stepped on the skin and the other only briefly looked at the hide as he walked up to it. Neither was frightened in the least by the bear skin. I had packed deer and elk for the first time on several llamas, and they always acted as if it was just another pack load. So I knew the llamas would be their usual tranquil selves on this new packing experience.

Several days earlier, while hunting in another part of Montana with friend Bob Garrott, we had found a large group of elk in a logged area, and Bob had shot the only legal bull in the group, a five-point. The next day, we hiked with the four llamas into the area. We packed the meat on three of the llamas and the other one carried all our personal gear. The packs on the llamas plowed six-inch furrows in the deep snow on each side of the llamas on the steep slope where the elk was shot. The llamas performed flawlessly as they followed one another back to our truck. Bob was so impressed that he asked that I call him if I shot a moose so that he could bring his wife and two-month old daughter to watch the llamas pack out the moose. After shooting the moose, I called Bob at his home near Quake Lake and made arrangements for him and his family to meet me at my camp the next morning.

LLAMAS ARE THE ULTIMATE

It's a good idea to use plenty of blaze orange when packing antlers or using llamas that are colored similar to the game being hunted. The llamas permitted us to get two nice bulls.

I had hauled the llamas from my home in Idaho Falls, Idaho to Montana in a stock trailer. However, I would not be able to get the stock trailer within 10 miles of where the moose carcass was located. So I would need to haul the llamas in the stock rack on the truck. That worried me a little. I hadn't owned llamas for very long so I usually backed the truck over to a hill or some other structure, and the llamas got in and out of the truck. However, there would certainly be no place to get off the path in the deep snow when we drove into the canyon below the moose carcass. It would have been difficult to back through the snow to a suitable loading place even in camp. Therefore, I decided that it would be best to make the llamas jump into the bed of the four-wheel drive. At that time only one of the llamas, Bandit, was trained to jump into the back of the truck. Although Chief Buck and Himalaya had been in the stock rack several times, they had never been

Larry Cadwell and Bandit pack out a five point bull elk in Montana. Where the elk was shot, the snow was deep, and the loaded panniers created six-inch deep furrows in the snow.

asked to jump into the truck from level ground. Churchill, the other llama, had been purchased earlier in the summer and had never been in a truck before. However, I knew he was a quick learner as all llamas are. We had purchased him in June when he was a three-year old. He had been caught once in his life prior to our two-hour effort to catch him. We worked with him several times after we got him home. We put the pack saddle on him one time while at our farm. The next time he had a pack on, he was on a two-day trek with us. The third time Churchill was saddled, he carried a deer for us, and on this trip, he had already carried the elk.

I shouldn't have worried. The llamas were staked out, each about 50 yards from the truck, in several directions. I brought Churchill in close to the back of the truck and tied him to a tree so he could watch the others load. Counting the time necessary to walk each llama to the truck, all were loaded within three to four minutes. Even Churchill who had never been in a stock rack jumped into the truck with little effort on his first try.

Bob's wife, Diane, led the four llamas in a string both going into and out of where the moose was located. Bob carried their infant daughter in a pack on his chest. Although it was Diane's first time to be around llamas, the two studs and two geldings behaved perfectly for her. After a brisk hike in the early morning cold, we arrived at the kill site. The llamas hardly noticed the bull as they settled down to eating the tops out of the small conifers sticking out of the snow and to which they were tied.

Llamas are becoming more popular as pack animals throughout North America. Their tranquil and gentle nature make them ideal for family trips and for packing game. As indicated I often take both studs and geldings on the same trip.

Not many other domestic pack animals will congenially pack in such combinations.

Llamas are ideal for people who may be intimidated by horses, or don't want the responsibility and expense of caring for horses all year or those who prefer to walk but do not want to pack heavy loads on their backs. Llamas also cause minimal environmental damage. When I follow my wife and her llamas on summer trips to the high county, her print is easier to see than the print of the llamas' feet. In soft soil or mud, the llamas leave a print similar to a large elk. At night, we stake out our llamas on a 20-foot cord. Their grazing impact is minimal; their pellets resemble those of deer droppings and can easily be spread with a kick of the foot. Because of their minimal impact, the Forest Service and Park Service have used llamas in areas with fragile environments.

A number of roads in the national forests have been closed to protect big-game herds. Several states report an increase in their elk and deer herds, which they claim is a result of these road closures. Closed roads mean more difficulty for hunters on foot, but they can take advantage of the good hunting areas this creates by packing in and camping in the back country. Llamas are ideal for this type of hunting.

Packing with llamas into high mountain lakes in the summer is truly enjoyable. They follow you effortlessly, in a pack string, anywhere you want to go. The llamas even carry our water canteens in the front of their panniers. We don't take freeze dried foods on these trips as my wife, Cherrie, packs one of our llamas with one or two ice chests. For example, on one trip in the summer, our first night meal consisted of beef tenderloin steaks and corn-on-the-cob cooked over a portable grill, baked potatoes, butter, sour cream, ice drinks and, of course, cherry cheese cake. The next morning, we enjoyed a leisurely breakfast of sausage and egg casserole, and bagels with pineapple and strawberry cream cheese, hot drinks and cold juice. Our remaining meals were also gourmet with salmon steaks and shrimp cocktail, and chicken and rice being the main courses for the next two night meals.

Another reason llamas are becoming more popular is that they are easily trained. The ease at which Churchill became accustomed to the saddle and panniers and jumped into the truck is typical of the llamas' ability to quickly learn new things and adapt to new situations. Llamas can even be taught to lie down in the back of a small airplane. Two of our packers recently were flown into Chamberlain Basin in the Frank Church River of No Return Wilderness and jet boated back up the Salmon River on the return trip out of the wilderness. I've also heard of a llama owner who taught his llama to lie down in a canoe. He paddles upstream to a remote area where he and the llama go deer hunting. If he scores, the llama carries the deer back to the canoe, and they float back downstream. Thus llamas can be flown into or packed upstream in boats into wilderness areas to assist in hunting big horn sheep and bear.

Early one fall, Cherrie and I were exploring an area in the Gravely Range in Montana with four llamas. Since this area was new to me, I frequently stopped and pulled out the topographic map and spread it out in front of me. Invariably, the stud immediately behind me walked up on the left side of me and looked over my arm, and the second llama in the string broke rank and peered over my right arm at the map. Cherrie suggested that they were just checking to make sure I was still heading in the right direction. Of course, llamas are intelligent animals, but I don't think they can read maps. However, this is another example as to why llamas are becoming more popular; llamas' intriguing and curious personalities make them fun to pack.

Llamas are calm on the trail. Once a llama has packed a few times, they just don't spook at anything. We have used some of our veteran packers to lead young or newly acquired untrained packers. The trailing llamas are tied to the back of the saddle of the llamas in front of them. On two occasions when we didn't tightened the cinches properly, failed to use the breast strap and were going up steep terrain, untrained llamas have pulled off the saddle and panniers from the llama in front of them in the pack string. The packs fell around their back legs. On both occasions, the llamas just stood, waiting for us to pick up their load and put it back on them.

One fall, two of us were on a hike with our llamas, and a friend was walking behind them, leading a lightly loaded horse. When the pack saddle on the horse began slipping, we were treated to a bucking spectacular for several minutes. The frightened horse would have easily scored 78 points at a rodeo. It eventually broke away from its owner and came near the llamas. The llama in the rear moved a few feet to avoid the bucking horse, but the llamas hardly gave the horse a second glance.

Shooting doesn't seem to bother llamas. When we first acquired llamas, I had planned to shoot some around them but just didn't get the time, so one day when the neighbors were going to shoot a .44 magnum pistol, I took a stud over behind them. He found some grass and began to eat. As the shots boomed, the llama ignored them and continued to eat his find of grass. The blasts hurt my ears, so we left; I didn't worry any more about their reaction to a gun being fired.

LLAMAS ARE THE ULTIMATE

I used to say that llamas are sure footed and can generally follow you anywhere you can go without using your hands. It's still true; however, we have discovered that they can go some places where we had to use our hands. Recently, we took our llamas over Hailey Pass in the Wind River Range in Wyoming. While going through a rock chute to get over the highest point, we had to use our hands to help lower ourselves three to four feet, dropping down from the boulders, while avoiding the ice covered portion of the chute. The llamas, all individually, were led down over the boulders and abruptly turned to avoid the ice chutes. We were later told by local individuals living in the community near the trailhead that we were the first individuals to take livestock over Hailey Pass before August first. Generally, those with horses avoid Hailey Pass because of this narrow, boulder covered chute at the top.

Llamas are also inexpensive to maintain. A bale of hay will generally feed a llama for eight to ten days. They are also relatively disease free. Almost any type of fence will keep them home. Llamas generally live 20 to 30 years.

Llamas do well in most climates. However, they are particularly well adapted to the cooler climates of the intermountain west. They do well in below zero weather. They just add more of their warm, luxurious fur in colder climates. At home my llamas have access to three-sided shelters which is all they need even in temperatures to 30 below. Their wool can be combed out or sheared and used for making beautiful clothes. Many women enjoy this extra benefit of owning llamas.

A mature pack animal generally weighs between 325 and 440 pounds. A male llama in good shape can carry 33% of his body weight. By boning portions of elk, we commonly use three llamas to carry out a mature elk or medium-sized moose.

PACKING GAME IS FUN 180

Mica, not yet two-years old, is led down over hugh boulders on steep Hailey Pass, Wind River Range. Most packers using horses avoid this pass. Below was an icy chute which complicated the decent.

For the archery hunter, a single male could carry out the largest deer if it were boned. Males generally mature when about three and half years old. They can pack smaller loads when they are younger. Generally for all day treks lasting several days, we try to load our llamas with about 20 to 25% of their body weight. If you need to carry heavier loads or want to ride, llamas may not be suitable for you.

Some horses are spooked by llamas. However, we have not had any problem with our llamas scaring horses. We generally leave the trail and warn on-coming people with horses that we have llamas and that they may scare their horses. Experienced and well-trained horses are usually not bothered much by llamas and other horses can adapt to being around llamas. However, horses have a reputation for being easily spooked so llama owners should give horses the right-of-way on the trail. The llamas are not at all frightened by the horses. Horse packers can get their horses accustomed to being around llamas in the off season so that their horses won't spook when meeting them out on the trail.

Llamas can be transported in a variety of vehicles. A mini pickup equipped with a stock rack can haul two or three llamas. I have seen a couple of big llamas transported in the back of a dodge van. Llamas usually lie down to travel, which lowers the center of gravity and helps vehicle stability. These particular llamas quickly got up and walked outside when the owner opened the side door. The owner indicated that they usually laid down for the entire trip, except the stud gets up when he recognizes the mountains near their home. Then he stands with his head between the seats, watching out the front window for his female llamas. These particular llamas were trained to relieve themselves outside the vehicle. People also have trained their llamas not to relieve themselves in their trucks or trailers.

Llamas can even be house broken. However, I understand that they are extremely hard on house plants. Their habit of depositing pellets in common locations in their corral or pasture is apparently helpful in house training. This same habit makes clean up of llama pens or pastures easy. Llamas do not have an odor, and their pellets can even be used as indoor plant fertilizer. I don't have llamas in my house, but I have taken a couple inside to see if they would climb the steep stairs to the basement. I give presentations to groups on llamas, and I wanted to be sure the llamas I use for demonstration would go upstairs to a conference room, if necessary. Both followed me up and down the stairs without any problem.

One of the most frequent questions I hear is, llamas spit, don't they? Yes, they do, and yes I have been spit on. I like to pet my female llamas, particularly when I feed then. Generally, I get spit on a few times each year because I am downwind from a couple of females squabbling over the best place at the hay manger. The other times we occasionally get spit on is when we bring a pregnant female into a stud for pregnancy testing. If she is pregnant, she wants nothing to do with the male so she tries to avoid him by rapid movements and by spitting toward him. They don't intentionally spit on me. Packing llamas rarely spit at anything on the trail. If you have packing males, you may go years without being spit on. Normal llamas never intentionally spit on people. The bits of vegetation in the spittle dry in a minute and can easily be wiped off. On the other hand, I don't get kicked or bitten by llamas; if a horse or mule kicks or bites you, you won't be able to wipe it off so simply.

Everyone should get in shape for hunting, and llamas can be good jogging companions. A friend of mine has run up to five miles at a time with one of our llamas; it helped him train for cross-country and road races. Llamas are fun to walk or run with as they relieve the boredom of running or walking by yourself.

Families enjoy llamas because they are relatively safe compared to other domestic pack stock. They usually do not kick or bite. One summer, we purchased a herd of 17 llamas to add to our existing herd. Some of them were two to four-years old and had either never been caught or caught only once. After placing a halter and lead on them, I pulled them forward into the trailer while Cherrie pushed by putting her hip on their rear. Neither of us were kicked. You can imagine what would have happen to Cherrie if she had pushed her hip into an untrained three-year old of other species of domestic stock. We maintain a herd of 55 to 70 llamas and are continually training and handling them, picking up their feet, etc. Cherrie has never been kicked, and I have only been kicked lightly a few times on the hand.

I've seen a friend's three-year-old daughter sit on the scattered hay in the middle of a group of feeding females. The females were as careful to avoid stepping on her as they fed all about her as if she were one of their own babies.

Perhaps the main reason you can expect to see more llamas packing game in the future is that llamas are just so much fun to own. They add considerably enjoyment to hunting and fishing trips. Even family members who do not hunt or fish seem to love llamas. The tranquility and joy of packing with an animal that silently follows you wherever you want to go is hard to explain. With their intelligent looking eyes, they quickly understand what is expected of them and go about doing it silently and without complaining. Their 6,000 years of packing experience in South America have bred into them a joy for the outdoors and their job requirements.

I no longer worry about how far from the road or what type of game animal is harvested. I know if I can climb to it without using my hands, my llama friends will be happy to visit the area with me and help haul out the critter. Isn't that what friends are supposed to do?

CHAPTER 25

HUNTING AND PACKING WITH LLAMAS

The sun was bright, especially on the snow, and the temperature was hovering just above the 10-degree mark. The nylon gaiters over my leather boots weren't designed for wading 12 inches of water. However, they did the job as I quickly made my way across the 20 foot of open water without getting water into by boots. Thankfully, I climbed up along a small tributary flowing into the water I had just crossed. Within a quarter of a mile, I spotted two deer moving among the aspen trees at the top of a sagebrush covered slope. Using a small draw, the only cover available, I quickly closed the distance to 300 yards, but the deer had disappeared into the thicker trees. Disappointed I lay in the snow for a while. Then the little buck reappeared, reversing his direction and walking through the aspen. The .243 was rock steady across a protruding snow covered rock as I held the crosshair at the top of his shoulders. As the buck dropped, I gleefully thought about packing him out.

We had hiked about three miles in from a road north of Kemereer, Wyoming to set up our camp the previous day. The buck was another mile from camp, but I was looking forward to packing out the deer. My daughter, Tammy, had harvested a buck the previous evening. My wife, Cherrie, had helped her field dress the deer and propped it up on a log to cool. Heck, I'd even pack that one at the same time which would add even more to our enjoyment.

No, thoughts like this don't make me crazy; for us, packing out game is the best part of a hunting trip! It wasn't always

HUNTING AND PACKING WITH LLAMAS 186

Llamas permit us to go to uncrowded areas where the hunting is best. During hunting seasons, our llamas often pack maximum loads.

that way. But on this trip, we had brought four friends, llamas all, which had packed in all our tents, equipment and food for the four hunters in our group. These remarkable animals have added so much year-round pleasure to our lives, but they are particularly fun during the hunting season.

The packing of the deer would be even more fun this time because one of the llamas was inexperienced and had never packed any game. That afternoon, Cherrie and I walked Himalaya and Churchill across the creek to where my deer was located. After loading the deer on Himalaya, we hiked with Himalaya and Churchill 30 minutes to where Tammy's two-point buck was located. We tied the llamas to nearby trees while we peeled the frozen skin off the deer. The llamas entertained themselves by watching a small bull moose as he walked through the opening in the trees. We cut the deer into two equal pieces, loaded each half in a game bag and dropped each into a pannier. Then the 120 pounds of deer and panniers were loaded on Churchill and the hide was rolled and tied on the top of the panniers. Although inexperienced and never having carried game, Churchill stood still and showed no reaction as we loaded the deer on him. The trip out was also uneventful.

We have packed other inexperienced llamas with deer, elk, moose, javelina and bear with similar results. To llamas, carrying out game is just another packing load, and after being a pack animal for 6,000 years in South America, they know their functions and responsibilities. I know that no matter what I put on a llama that has packed a couple of times, he'll carry it without complaint. None-the-less, every time I put a game animal on a first-time llama, I am thrilled by their "It's no big deal attitude."

Churchill and Tammy Parker pack over 130 pounds of deer and supplies. On this his second trek, Churchill paid little attention to our skinning her deer or putting it in the panniers on his back.

LLAMAS ARE THE ULTIMATE 189

Dave Lechel leads Bandit with a desert mule deer and Jasper with a javelina in southwestern New Mexico. Jasper had packed only once previous. The javelina didn't bother him at all.

Llamas are sure footed and can go just about anywhere a person can without using his hands. Several years ago my hunting buddy, Don Martin, and I packed out our first elk with llamas. We had scored on a couple of nice bulls in the Grays River area east of Alpine, Wyoming. We had shot the bulls not too far from our camp which was about four miles from the road. The snow had begun to melt and the trails were slick and muddy. In some of the steeper places, the horses had slid and gouged out the trail. We had slid down several times ourselves on the two trips to carry out our camp and elk with the three llamas. However, our three sure-footed companions didn't even slip although each was carrying 95 to 125 pounds.

HUNTING AND PACKING WITH LLAMAS

Llamas have two soft pads on each foot which help them climb. Their tracks resemble over-sized elk tracks. However, I can usually see the foot prints of my wife easier than I can the llamas' prints when I follow her on a summer packing trip. This sure footedness and lack of environmental disturbance, is one of the reasons that the Forest Service and the Park Service use llamas to pack in fragile environments.

Archery hunters are discovering that llamas, with their silent walking and calm behavior, make excellent hunting companions. A large deer can be carried out with one llama if the meat is boned.

Llamas can carry a maximum of about 33% of their body weight. Most of our packing males weigh over 400 pounds. So we use three llamas to carry out a mature bull elk or a young moose. One llama can carry out the largest deer if the deer is boned. Archers are finding llamas ideal for bowhunting. They can silently follow a bow hunter. Several bowhunters, including Dwight Schuh, a prominent author and lecturer, have purchased llamas from us. All of them have expressed satisfaction about how their llamas have increased their hunting enjoyment. For trips lasting several days, when the llamas pack for a long time each day, we like to load them with 20 to 25% of their body weight. For example, on a trip into the Cloud Peak Wilderness Area in the Big Horn national Forest in north central Wyoming, we packed our llamas with about 85 pounds each and covered 25 miles in one weekend. They made the trip much easier than we did.

One of the reasons llamas are catching on in North America is that they are so easy to train. Recently we taught, in less than an hour and a half, five baby llamas between five and six-months old to jump into the back of our trailer which is two foot off the ground. During the first three times each llama was loaded, we had to tug hard on the lead rope and lift on each youngster to get him or her into the trailer. On the forth try, each llama was loaded by only pulling on the lead rope without any lifting. On the fifth try, each baby llama easily jumped into the trailer. Llama owners like to say that if you show a llama something three times that he will learn it. I believe it. After we finished our training session, we loaded a young female and tied her in the trailer in order to transport her for a health certificate examination. We left for a few minutes with the trailer door open. When we returned, one of the llamas we had just trained and released had jumped into the trailer. So the experience apparently wasn't very traumatic.

HUNTING AND PACKING WITH LLAMAS

Llamas are intelligent. Not only do they learn quickly, but they retain what they have learned. My wife and I sacrificed our packing trips into the high country one previous summer so that we could construct fences and facilities on a small farm we purchased for our llamas. Therefore, we didn't get an opportunity to take any of our llamas packing. That fall when we started hunting in Wyoming, the llamas had not packed in a year. We packed seven llamas for an elk hunt. From their behavior on the trail, one would have thought they had been packed all summer and fall. Even a three-year old, Bandalero, who had a saddle and empty panniers on him once in 1987 and once in 1988, packed like a veteran.

Once llamas have been on the trail a few times, they gain confidence in you and will follow almost anywhere. Late one summer, we hiked into the Jebidiah Smith Wilderness Area adjacent to the west side of the Tetons. On the way out, we took a different route which took much longer to travel than the way into the area. We approached the camping area where we had left our truck at about midnight. There was no moon and it was so cloudy that visibility was very limited, and our flashlights had quit working about an hour before. I followed the noise from our dog's tags to negotiate the switch backs on the last mile of the trail. Cherrie had stayed on the trail by putting her hand on the rump of the llama I was leading. We were dead tired and did not want to wade the creek so we decided to cross an 18-inch wide walking bridge across the stream. Without even breaking stride, the llamas followed us across the tiny bridge. They will even follow you into a building and climb up and down stairs.

Another reason for llamas gaining in popularity is that they are the safest pack animal being used in the west. It is unusual for a llama to kick or bite. We routinely handle 60 or more llamas on our farm. I have never been bitten and have been tapped on the wrist twice while combing the back legs of a couple of young males.

Also hunters and summer hikers are finding that llamas on the trail seldom spook at anything. Hides and antlers tied to the tops of panniers aren't a problem for llamas even when they drag the ground or rub on trees.

When we camp with llamas, we tie them with a 25-foot cord which allows them room to graze. After being tied a few times, it is rare for a llama to tangle his feet or legs in the cord. They continually watch for the cord and step over it. I know if they tangle their cord in bushes or on rocks, they will not panic. They will simply lie down and wait for us to come to their aid.

Another reason for sportsmen turning to llamas is their appeal to the entire family. The wool can be collected, spun and made into beautiful clothes. Women, in particular, fall instantly in love with llamas. We have had many outdoorsmen come to our farm who were considering purchasing a llama. Sometimes the spouse is reluctant to purchase a pack animal before coming to the farm. Invariably, the spouse falls in love with the llama, and often is the most enthusiastic about purchasing one or two of them before they leave the farm. Because of their gentle nature and easy trainability and temperament, they also have appeal to the family as 4-H projects. Parents don't have to be so concerned about their children getting hurt by a llama. Llamas look cuddly and their large eyes with their over-sized eyelashes can melt the strongest resistance.

This appeal at purchase time seems to last with the whole family. We have found that in many of the couples, one individual previously did not do any off road camping. But

HUNTING AND PACKING WITH LLAMAS

after purchasing llamas both now look forward to packing them into the wilderness and other roadless areas. This has been a particularly thrilling aspect of owning llamas for many couples. Cherrie and I never camped away from the road before we purchased llamas. Now we both are enthusiastic wilderness trekkers. In the summer time, Cherrie equips one of the llamas with two large ice chests. From these chests we enjoy such meals as filet minion, baked potatoes, corn on the cob, cold drinks and cherry cheese cake or a meal of shrimp cocktail, salmon steaks and asparagus spears. This isn't exactly roughing it. Recently five of us hiked 58 miles across the Wind River Range of Wyoming. We packed four ice chests full of gourmet foods and enjoyed some wonderful dutch oven cooking. Every night the dutch ovens were filled with delicious foods, for example pork chops and rice, marinated chicken breasts, rack of lamb, corn bread, biscuits, baked potatoes, elk stew, birthday cake, scalloped potatoes and apple crisp. Other foods were cooked on the grill, such as steaks, or on portable light weight stoves, such as corn and other vegetables.

Another reason for their appeal is that adult llamas can be hauled in almost any type of vehicle with the exception of a sedan. I've seen a number of llamas hauled in vans and minivans. We provide a stud service and a customer from Montana brings adult females to Idaho Falls in the back of his new minivan. A van will easily haul two adult llamas. Two also can be hauled in a mini pickup, and three large adults can be carried in the back of a full-sized pickup. For short hauls, four can be carried in a full-sized pickup. However, llamas like to lay down when they travel so room needs to be provided for them to lie down on long trips. Their lying down improves the handling of the vehicle as it lowers the center of gravity. Llama are versatile riders. They can be taught to lie down and ride in a boat or canoe. They can even learn to ride in the back of a single engine airplane.

Llamas are also economical to own. A bale of hay will feed an adult llama for seven to ten days. Before we purchased our farm, we had rented a pasture from a farmer who indicated that eight cows would eat all the grass in the pasture. We put 25 of our females and babies in the pasture and the grass eventually seeded because the llamas couldn't eat it fast enough.

Almost any type of fence will keep a llama home and happy. However, it should be pointed out that llamas do not adjust well to a solitary existence. A three-sided shelter is all that we use to provide protection against the wind even in our Idaho Falls environment where temperatures to 30 below in January are not uncommon. The llamas just put on more luxurious wool.

Llamas are not expensive especially when their useful life is considered. We know a lady who put a llama into her commercial pack string when he was 15-years old and packed with him until he was 27. Prices for male llamas normally range from about $1000 to $2500 (US dollars).

Llamas are curious. When we start a project on our farm, a group of llamas normally gather around to check out the activity. Similarly, when a new baby is born, all the llamas in the pasture gather around to inspect the new arrival. Llamas also seem to enjoy the back country trips as much as we do as they continually look around to see the sights. Their intelligence, their no nonsense approach to packing and their gentle and interesting personalities all help add to the enjoyment of owning llamas. However, the best reason for owning llamas is that they add so much enjoyment to summer trips to high mountain lakes and to fall hunting trips.

CHAPTER 26

BACK COUNTRY FISHING WITH LLAMAS

Out of the corner of my eye, I noticed that Les had also seen the big trout break the water surface for the insect that had landed on the water. The trout shimmered as he completed his roll back beneath the surface. Les had already sent his fly arching through the calm toward the widening circles on the water before I could even reel in my lure. Since I hadn't hooked a fish in half an hour, I sat down to admire the scenery.

Before my eyes was a scene that could have adorned the cover of a travel magazine: a high mountain lake nestled within snow covered granite peaks with fluffy white clouds suspended in an unpolluted blue sky.

Although fishing certainly adds interest, these high mountain lakes are situated in remarkably beautiful country. Often these lakes are found just beneath the highest peaks where former glazers have gouged out a lake, and there is plenty of snow pack to replenish the crystal clear water. These lakes usually occur in both wilderness and nonwilderness roadless areas and are normally not crowded. Back country lakes are often reachable only by foot or with pack stock.

Although many backpack into these gorgeous surroundings to sample the fishing and to breath in the cool, clean air and view the specular sights, back packing requires a major effort to reach the highest country. Cherrie and I seldom ventured that far from our vehicle, that is before we owned llamas. Llamas have opened up a whole new world for us! The

The high country lakes in the wilderness and roadless areas of the western United States and Canada are located in the most magnificent country in the world.

wilderness and other roadless areas provide, without a doubt, some of the most spectacular scenery in all of North America! The pristine lakes usually associated with the highest peaks also provide unique fishing opportunities. In the west, the trails into the lakes often go through several types of habitat from desert floor to alpine meadows in only a few miles. Similarly, in other parts of North America, llamas provide easy access to uncrowded vistas and fishing. Thus, the trails lead to some beautiful country. Photo opportunities of the awesome peaks in the back ground usually abound on the trails.

Surprisingly, many of these lakes are only a few miles from the trail head. Therefore, many lakes are tailor-made for weekend llama treks. Connecting trails often lead to other mountain lakes. A three to seven-day llama trek can sample several or many lakes. Where we live in southeastern Idaho, there are more high country lakes and trails in Idaho and adjacent Wyoming and Montana than we can possible visit in a life time. Opportunities for summer llama trekking in much of North America are unlimited.

Llamas carry all the necessary fishing equipment. Usually float tubes and small rubber boats are deflated and placed in a pannier or tied on top for the trek into the lakes. However, if the trail is wide enough to assure that a limb won't damage the float tubes, they can be carried inflated on top of the panniers. Fishing rods can be placed in tubes and tied to the top of the panniers. Tackle and fly boxes are usually placed in the panniers.

Although simple meals taste great on these fishing trips into the high country, summers are the time to pack in gourmet meals. There's plenty of time for cooking and plenty of room for frozen meats, vegetables and desserts. Ice chests packed with frozen and fresh foods and cold drinks make meals delightful and as gourmet as you choose. Cherrie often has one or two 30-quart ice chest in our panniers stuffed with some amazingly good grub. If you decide to go without the ice chest, drinks can be cooled in a nearby snow drift or cold mountain stream. A roaring camp fire, delicious meals, good companionship and good fishing while viewing some of God's finest handy work makes llama treks in the summer wonderful experiences.

I'm an avid hunter and most of my free lance writing is about hunting. However, in my opinion if you own pack llamas,

If the trail is wide, float tubes can be carried inflated.

their highest and best use is packing into the high country in the summer.

In a previous chapter, I mentioned that often couples begin packing into the wilderness together after purchasing llamas. Hiking into the wilderness in the summer with llamas is an experience that couples particularly enjoy doing together. Llamas make it easy for the entire family to visit high mountain lakes.

Vacations or summer llama treks can be something that you will long remember. For example, in the Wind River Range of Wyoming, you can choose trail systems that take you from a few miles to over a 100 miles with dozens of lakes to fish, permanent glazers, wildlife and wild flowers galore. Places like this exist in many states and provinces, but few people take advantage of them. However, llamas make these types of vacations fun, economical and achievable.

As I looked at the spray looping down from a water fall, fed by the melting snow, my view passed over the top of a bent, pulsating rod tip. Les had hooked a fish in the same spot where the hugh trout had surfaced. What luck I thought. A 13-inch cutthroat suddenly danced on the surface. It certainly wasn't the big fish that had taken the insect. Les had already caught several fish up to 20-inches long, but he was hoping for another big one. The three of us had caught and released a number of fish from this lake.

Sunset found the three of us sitting around the camp fire as filet mignons and corn on the cob sizzled on the grill above the fire, and potatoes baked in the coals. Later, hot cherry pies would be made and cooked over the coals. The warmth of the fire was welcome as the nights can be cold at these back country lakes even in early August. Although the days are normally pleasant and comfortable in mid summer, a jacket may be necessary even in mid day. Warm sleeping bags are often a necessity at the highest lakes as you can wake up to frost-covered vegetation or a brief early or late summer snow storm.

The next morning, we fished a little while and then loaded the llamas for a short half-mile hike to the next lake. Although we fished only two lakes on this trip, a total of five lakes existed in this particular area, each separated by about a half mile, making a five lake weekend fishing trek possible. One of the lakes has golden trout, at least two of them have a combination of cutthroats and rainbows and the others contain cutthroat trout.

BACK COUNTRY FISHING WITH LLAMAS

Cherrie pulls in a trout from a lake in a beautiful setting. Fishing is just part of the fun in visiting high mountain lakes.

July and August are usually the best llama trekking months at the higher elevations. Depending upon the winter snow pack and spring temperatures, some lakes are often still frozen in early June, and the trails may be blocked by hugh snow drifts until late June. September is often the best fishing month and has fewer insects. However, fishing treks with llamas is not limited to the summer months as even in western North America, trails are accessible into wilderness and roadless areas of the lower elevations in late winter and early spring. The trails often parallel rivers or streams. In the spring, wildlife concentrate in these areas to escape the deep snow at higher elevations, making excellent opportunities for wildlife viewing and photography.

Usually there is plenty of lush grazing for the llamas near the high mountain lakes. The llamas are easy to care for as all that is required is to change their grazing location a few times each day and to take them to water once a day.

Of course, fishing can be great or poor at these remote lakes and streams, depending upon conditions and the lake selected. Some lakes are known for their large fish and the difficulty in catching them. The size of fish varies between lakes. However, many lakes are very productive and produce lots of nice fish. Eating a freshly caught trout along with your other breakfast foods is a real treat for some folks. A platter full of freshly grilled trout with your favorite trimmings for dinner is always a welcome treat. However, with llamas it's easy to bring along other delicious foods in case the fish don't cooperate.

For a vacation that features beautiful scenery, solitude and good fishing, a llama trek into a scenic high mountain lake is hard to beat. In the spring, a trek into a lower river drainage can be a wildlife viewing spectacular. For me, fishing and

Many lakes in the high country are only a few miles from the trail head.

vista viewing in the mountains may be the highest and best use of llamas. It's an experience that gets you away from it all. Its economical, fun and healthy both for the mind and the body. The key ingredient for all this to happen is owning llamas!

Bandit admires the two-pound mackinaw caught in a alpine lake. Wow, were they ever delicious!

CHAPTER 27

LLAMAS, THE ULTIMATE SHEEP GUARDS

Sheep are one of the most common livestock raised throughout the United States. They exist in flocks of a few sheep raised in farm flocks to thousands on hugh ranches and public lands of the west. Sheep are sometimes vulnerable to weather and other environmental purls, diseases and predators. The lambs, in particular, are easy prey for coyotes.

Coyotes are an important, natural component of the environment. They have maintained viable populations and even expanded their range despite continued attempts at eradication. Coyotes are an important part of many natural ecosystems; however, sheep owners who suffer losses or can even be put out of business by coyotes, find it difficult to maintain a positive attitude about them.

Dogs also kill sheep. Losses to dogs can be locally severe on small farms in close proximity to human dwellings. The offending dogs are almost impossible to apprehend or to destroy. Like coyotes, dogs often strike at night and are extremely sensitive to human approach.

Since I live in Idaho Falls, Idaho most of my contacts are with western llama and sheep breeders. In addition 80% of the sheep in the United States and 39% of the operators live in the 17 western states. Data are mostly available for western flocks since nearly all federal efforts to control coyotes and to study predation takes place on public lands in the west. However, similar problems and solutions exist throughout North America.

Sheep predation is a substantial and important economic problem. Studies show the average annual loss rate to predators ranges from 4 to 8% of the lamb crop and 1 to 2.5% of the ewes with 76 to 100% of the kills due to coyotes. In 1978 an estimated 19 to 38 million dollars worth of sheep were lost to predators just in the 17 western states. This is not likely the full loss since predators raiding a flock undoubtedly cause indirect effects such as decline in weight gain and lambing problems. In addition, the annual cost of predator control was estimated to be over 14 million nation wide.

Although one western study indicated predator losses accounted for less than half of the total deaths due to all causes, predators can easily account for the difference between profit and loss and survival in the sheep industry. In the 17 western states, the sheep industry has declined from 68 thousand operators in 1968 to 47 thousand operators in 1978 and a reduction from 15 million sheep to 8.7 million. This decline was due to a number of factors, including price for mutton and wool and other economic conditions, lack of competent laborers and herders, advancing age of the operators, and, of course, predation. A total of 10.8 million sheep in the United States as of January 1, 1989 indicates some stability since 1978.

Many criticize current coyote control because it is non-selective, that is many coyotes that do not prey on sheep are destroyed. Since 1972, 1080 and other poisons are no longer widely used on public land because they affected other species of wildlife. Therefore, current common methods of coyote control are gunning from planes, denning, that is finding and destroying pups in the den, and trapping. In addition, the ancient practice of using dogs as guardians has been revived. However, llamas can provide an attractive alternative to prevent sheep predation. In fact llamas may be the most economical and effective method available!

Llamas are excellent sheep guards, usually 100% effective. This 20-month old llama almost knocked over a fence and four posts trying to protect a lamb attacked by a dog across the fence.

Although llamas are gaining quite a reputation for being easily trained, tranquil pack animals and excellent investments, not many people are aware of their sheep guarding abilities. Recently I interviewed a number of sheep producers who have used llamas for this purpose.

"That first year, we had a real snow storm in September. They were caught out away from the buildings during the night when the storm hit. We had to fight our way to the sheep so they could come home in our tracks. The llama never, never let those sheep get caught in another storm. If it were to storm during the night, he would have all the sheep in the shed the next morning."

Eldora Porch and her husband were major sheep producers. However, they were forced to sell most of them one year when coyotes killed over 100 lambs in spite of the sheep being herded into pens each night. In the next few years, they lost five or six per year out of a flock of 40 even with all their precautions. Four years ago, they purchased a llama. They have not lost a single sheep to predators since then, even with the sheep remaining out at night in large, rough sagebrush pastures.

These are only two of the amazing stories related by sheep breeders. In fact, every individual, without exception, praised their sheep guarding llama. Common comments were, "We are just as pleased with him as we can be." or "We just can't sing enough praises for that fellow. He is probably the best investment we have ever made." or "He is the only animal on the place that is insured. Why we would loose a bunch of money if something happened to him." or "He is irreplaceable" or "He's worth his weight in gold." or "We paid a thousand for him; I sure wouldn't take that for him now." or "I'm darn proud of that old boy; he's a smart critter."

It is easy to understand why these people are so positive about llamas for guarding sheep. Several said their llama had saved them thousands of dollars each year. This behavior of guarding and protecting was inherited from their ancestors. This is the reason why sheep guarding comes "naturally" and doesn't requires training. Llamas were developed by selective breeding of guanacos by South Americans 4,500 to 6,000

A llama keeps the sheep in a tight flock. Soon the sheep begin to depend on the llama for protection and guidance.

years ago. Guanaco studs commonly collect a harem of females and defend territories.

Most llamas were between one and six-years old when they were introduced to sheep flocks. Although llamas do not reach maturity until they are three to four-years old, immature llamas were surprisingly effective. Only two one-year old llamas were not effective. One was successful his second year; one was exchanged for another llama two weeks after purchase because he preferred to be with horses that were pastured with the sheep. Three sheep breeders put nine-month old llamas in with their sheep. Two stopped coyote losses almost immediately, and the third one eliminated predation when he was one and one fourth-years old. However, the young llamas became more aggressive toward predators as they maturated.

Nearly all of the llamas guarding sheep were geldings. Intact males may injure or kill ewes when the ewes are in season. One llama per flock was recommended. One operator put three llamas into one flock, and they bonded to each other rather than to the sheep.

Llamas are extremely effective in preventing coyote and dog losses in both large and small pastures. Curt Reisland runs sheep on 600 to 2,000 acre pastures. He had predator losses of 50 to 60 lambs per year before March 1987 when he purchased a yearling llama and has not lost a single lamb to predators since then.

Coyotes killed sheep within 25 yards of Harold Hanson's home, and dogs and coyotes took one to three lambs every year in his small pasture. In the three and half years since he purchased a llama, he hasn't lost a lamb.

Before getting a llama, George Richardson lost 25 to 30 lambs per year to coyotes. During the three years that his llama has run with his 500 ewes, he has not lost a single sheep to predators. He has his sheep in rough country at the head of a drainage.

In three years, Charolette and Ralph Carlat have not lost a ewe or lamb to coyotes when the llama was with one of their flocks. Once they moved a flock of 200 ewes and their lambs into a 1200 acre pasture without the llama. Immediately they lost four or five lambs to coyotes. Their llama was taken to this pasture, and coyote losses ceased.

Six ewes and nine lambs were taken by coyotes during the first year Ross Middlemist had a yearling llama. During the next two years, he didn't loose any on his 30-acre enclosure surrounded by open range. He and his neighbors have seen the llama chase coyotes.

Coyotes and grizzly bears took 100 lambs per year on Dallen Smith's 2.5 sections in Canada before he purchased llamas. He has had an annual loss of about 20 lambs in the 2.5 years after the llamas were introduced. One night, sheep were put in a small electrified corral without the llamas, and coyotes went through the electric fence and killed several.

Patty Bailey had three lambs killed by coyotes before purchasing a llama; afterwards, sheep losses in her 40-acre pasture ceased.

During the last three years with a llama, Skip Hurt hasn't lost a single sheep to dogs. His next door neighbors, who have similar sheep operations, experienced dog kills every year; one lost 22 lambs in 1988 alone.

John Lyle cut his losses from a high of 36 to 40 lambs per year, 8 to 10% of his lamb crop, to zero losses during the next five years after putting a llama with his flock in 100-acre pastures.

According to the producers, just before or at lambing is the ideal time to place a llama with sheep as they immediately bond to new-born lambs. "He loves the lambs" or "Crazy about the lambs" or "He just watches over them like crazy," were common statements from every producer interviewed. One breeder indicated that his llama watches within two feet of where lambs are being born. Several others reported that their newly introduced llamas began to "Moan or whine" when herded away from the sheep and lambs.

Another had a yearling llama in with her herd for two weeks, "The llama was standing out there in the pasture by himself and wouldn't move. My husband went out there, and he was guarding a sick lamb." A week later the llama was out on a plateau by himself all day. "When we went out to the area, a ewe had given birth to lambs, and he was just standing there over the lambs."

Another breeder introduced a llama at lambing, "We didn't know what he would do so I turned him in with the yearling bucks just to get him used to sheep. We had little lambs on the other side of the fence. He kept walking the fence. We turned him in with the lambs, and he would lay in the middle of them, and they would come and play with him and jump on him."

Another introduced his llama into the sheep four or five weeks before lambing, but the llama did not take up with the sheep until the lambs were born. "The first set of lambs that were born up there, I didn't know what was happening. He was stewing around more than the ewe. I was going to pack them in, but he was making the funniest racket. He just went nuts for the lambs when he saw them."

The guard llamas keep a close watch on new-born lambs. "The llama was lying down, and the lambs were using him as a spring board when a yearling ewe started playing with another group of llamas over in the corner of the corral. The llama got up and cut out the yearling ewe and then returned to let the other lambs continue to jump off him."

"He just loves the lambs. When I drive sheep out to pasture after we take lambs off them, what should be a 30-minute job takes an hour and a half. He wants to stay with the lambs."

"In January 1988, we started to lamb early. I looked out the window, and the llama was lying down on a hill. When we drove up there, the ewe had dropped twins, and the llama was laying between the lambs and the wind, keeping the wind off them."

Several reported seeing their llama in action against coyotes and dogs. "A coyote was more or less circling the flock, and the ewes were gathered right around the llama. In fact, there were some of them almost under him, and they were watching the coyote. The llama was just zeroed in - eye balling him."

"A neighbor saw a coyote come in and run two ewes away from the flock. He got out of the neighbor's sight with the ewes just as the llama took off. The next thing he saw was the coyote running away and the llama coming back with the ewes."

"There was a hell of a commotion. I looked out and the llama had the sheep backed into a corner, and he was taking on two Siberian huskies, wheeling and striking at them."

Others reported visitors letting their dogs out near their sheep. The llamas commonly chased the dogs, often back into the truck. "The llama just put his head down and started sweeping his neck back and forth and charged the dog and striking with his feet."

One breeder reported that when his llama sees a stray dog in the pasture, "He chases the sheep into the pen and goes to the pen gate and watches the dog. If they get too close, he runs after them and chases them away".

Llamas do more than just protect the sheep from predators. In a large pasture: "He came up by himself and came as close to the buildings as he could and paced up and down the fence. He would look over the hill like he was pointing at something and I said to my husband, 'You know he is trying to tell us something.' Sure enough we drove over to the pasture, and a couple of sheep had gotten stuck in a reservoir."

"I was standing in the garden when he came up. He reached his head through the fence and hummed. I went to the tank and something had knocked the drain pipe out, and the llama and sheep were without water."

"We had kind of a hay shed down there and every night at dusk, the llama would bring the 84 ewes into the shed or into the corral by the shed. Then he would stand in front of them. You couldn't get the sheep out. Right at sunrise, he would let them out."

Although several had not seen their llamas chase coyotes or dogs, they have witnessed their llamas control the flock. "The trapper snared a coyote that summer, and the llama would not let the sheep go through the open gate to that 1,200-acre pasture until the coyote was removed. The coyote was quite a ways from the fence or gate."

Soon after a llama was placed in a flock, "We had a stray tomcat run across the pasture. The llama chased the cat down into the creek. He spent about five minutes down there. I later looked, and he was heading the sheep for the far corner of the pasture."

An experienced llama was placed into a flock which had recent coyote kills, "The sheep were in little groups and scattered out. The llama rounded them up; he did this all on his own. It took him a few days to get them all rounded into one group. From then on, they were never separated from the llama. When they bedded down at night, he was right with them in a tight group. When they were eating, he would be on a higher level so he could see them because they spread out some while eating."

Sheep soon learn to depend upon the llama. Several indicated that the llama takes a leadership role even when the sheep are being herded. "As long as the llama was in the lead, the sheep easily went through gates. He got around behind one time and at the next gate, none of them went through. We had a dog with us, but none of them went through until the llama did."

Producers move llama guarded sheep successfully with the aid of dogs. There wasn't a problem unless the dogs push the sheep too fast, then the llama "Took after the dogs when they were chasing the lambs, and the lambs were squalling trying to get back to their mothers." or "Goes after them with his head real low to the ground and his ears back. Our dogs know its time to get out of there."

Llamas and sheep can be profitably raised together. My wife and I purchased sheep to go with our llama operation. Without predator problems, sheep provide a welcome addition to our income. Floyd Colpitts had severe predator losses for several years. However, when he lost 26 out of 79 lambs one year to coyotes, he was forced to sell his sheep. The next year he began breeding llamas and also purchased sheep. Although coyotes are still prevalent as he has seen his llamas chase them, he hasn't lost a lamb to coyotes in 10 years. He often has a combination of llamas, including females, studs and babies in with the sheep flock.

Although llamas are not aggressive toward people, several separate the llama when they handle the sheep, and

LLAMAS ARE THE ULTIMATE

particularly the lambs. Some don't want to upset their llama while others are apprehensive about what the llama might do. One owner stated that when shearers grab the sheep, his llama chases the sheep into a corner and stands between the sheep and the shearers.

For nearly all of the interviewed sheep producers, llamas completely stopped predator losses. Also llamas provided other benefits to some of the flocks such as protecting the sheep from severe weather. Although preventing predation is the most important economic factor, many of their characteristics help make them affordable even for small scale sheep breeders. Llamas live 20 years or more, providing a long profitable return on the original $1,000 to $2,000 investment. Food consumption is minimal; a bale of hay will feed a llama eight to ten days. They require no special diet since they eat the same things sheep do. They are normally gentle and tranquil; llamas seldom get excited. They require little health care, and they are hardy animals that can survive in any adverse weather where you would expect sheep to be raised even in Idaho Falls, Idaho, where we had temperatures of -30 with -70 degree wind chill during the winter of 1988-89. Since they are so easy to train, they also can serve as pack animals, 4H projects or produce wool for that special sweater. Their intelligence, extreme curiosity and interest in everything around them, help make llamas interesting and entertaining. It's easy to see why sheep producers are beginning to depend upon these remarkable sheep guards. Another reason for having llama sheep guards is their attractiveness. Have you ever seen a sheep band that wouldn't have looked better with a llama in it? Neither have I.

CHAPTER 28

FLYING & JET BOATING WITH LLAMAS

Mike Tupper suddenly stopped as the buzzzz filled the air. I was abruptly halted by the rump of his llama, War Bonnet. Himalaya and Bandit also bumped against me as we all slammed to a stop. The buzzing rattler kept dancing out his tune somewhere just ahead on the narrow trail chocked on both sides by a thick canopy of last year's grass. The trail clung to the side of the steep slope. A couple of years previous near this spot, two pack horses had slipped on late fall ice and had fallen to their deaths so there was no way to proceed except down the trail, and the rattler had no intention of leaving his grassy hide away.

The llamas weren't at all concerned as they twisted their long necks to the side to feed on the luxurious growth. However, we needed to go down the path so Mike cautiously peered in and under the thick grass. The buzzz picked up in intensity.

Mike reached back into the right pannier on War Bonnet and fetched his 44-magnum pistol. All three llamas continued to feed with their heads buried in the tall grass. I held my hands over my ears as Mike took deliberate aim as he crotched to look under the grass. The loud obtrusive roar even drowned out the consistent loud crescendo of Chamberlain Creek made powerful by a quick spring runoff. At the shot, one of the llamas calmly raised his head as if to indicate that these were tasty morsels; the other two had no reaction! I wasn't surprised.

LLAMAS ARE THE ULTIMATE

Later, Mike, who has experience with other types of pack stock, admitted that he would not even have considered shooting unexpectedly and especially that close to pack horses. He speculated that their reaction would have been either to run over us or to have jumped around and fell off the cliff or to have run the 18 miles back to our starting point. Of course, if we had been restricted to horses or mules, we wouldn't have flown them into the wilderness or be on our way down the steep trail to meet a jet boat for a ride up the Salmon River to a road. But I'm ahead of the events in this story.

I'm not sure when planning for this adventure really started, but a couple of years earlier, Mike and I had discussed the possibilities of flying with llamas. The next fall, I ran into Mike, and he was constructing a training box and planned to use it as an aid to teach his llamas how to enter his airplane. He wondered if I might be interested in flying some llamas into the wilderness, and we might consider riding back up the Salmon River in a jet boat. Of course, after no consideration at all, I readily agreed to come up the next spring as soon as the snow left the high country landing strips.

In late April, I loaded two of my llamas, Bandit a 410-pound stud and Himalaya, a 365-pound gelding, and began the four-hour drive to Salmon, Idaho. Upon arrival at Mike's home, he showed me how he had taught his gelding, War Bonnet, to enter a 7 foot long x 36 inch wide by 40 inch high box with a 36 inch x 40 inch opening. The bottom of the box was suspended 24 inches off the ground by 2 x 4s. Mike explained how if a llama could crawl into the dark box, he should be able to get through the larger opening into the larger passenger/freight compartment of his Cessna 206 single-engine airplane. I couldn't argue with that theory.

FLYING & JET BOATING WITH LLAMAS

Himalaya is loaded into the airplane for the trip to Chamberlain Basin in the Wilderness of No Return.

Then it was my turn. I took Himalaya to the box. I'm 6' 3" tall so I had to crawl into the box on my knees, turn my big carcass around to face the opening, then back into the rear of the box. By the time I had accomplished this task, Himalaya had bent his legs, placed his knees on the floor of the box and had begun to crawl in after me. In no time at all, he was in the box and staring at me face to face. I suppose he thought, "Boss you have really gotten us into another fine mess; there is no place to go in this contraption." If you haven't seen a big llama turn around on his knees in a 36-inch wide space, you can't appreciate how agile they really are; Himalaya easily made the turn just after I had slipped by him in the narrow closet.

Well the first one was certainly easy, now for the big stud. Although Bandit took more encouragement and a little shoving, we were able to load him in only a few minutes. On the third try, we still had to do a little pushing to get his big rear around the corner and into the box. The total time for the entire session beginning with putting War Bonnet in the box and ending with Bandit was less than 30 minutes. We were in a hurry to get into the wilderness so we decided that was enough training and it was time to try them on the real thing.

We quickly loaded our gear and the three llamas into the trailer and headed for the airport. We eyed the thick snow clouds as they whirled above us, and wondered if they would permit us to fly. As we traveled along, Mike indicated that it looked a little better to the northwest where we would be flying, and he thought we could make it. I was apprehensive as I couldn't tell that the clouds were any thinner in that direction than any other. I immediately thought of the questions my friends had asked when they learned I was going to fly llamas into the wilderness. "Is the fellow you are flying with a good pilot?" and "What kind of a plane does he own?" Admittedly, my answer to both questions was, "I don't know." They shook their heads and said, "You're crazy." Wow, those clouds did look thick, maybe they were right! Of course, when we later flew out of the Salmon River valley and headed up over the mountains, the clouds began to quickly dissipate. Later, I learned that Mike was indeed a safe and experienced pilot who has landed many times on remote wilderness airstrips.

Llamas are tranquil creatures but when you are flying over continuous peaks and trees, you don't want a llama climbing into the cockpit to help with the flying. So in order to make llama flying completely safe, Mike designed a llama safety harness or seat belt. The design takes advantage of the rear seat latches on the floor of the cargo compartment after the rear seats are removed. The harness consists of an adjustable collar that fastens around the base of the neck. This collar is

FLYING & JET BOATING WITH LLAMAS

A llama looking out the window as we fly by the Bighorn Craigs.

attached to a one-inch wide leather strap that goes along the back of the llama all the way to just above the hips. Attached to this back strap is a cinch belt that goes around the llama just in back of the front legs. At the hips where the back strap ends, a 12-inch leather strap attaches at right angles and hangs over each hip. Attached to each side of the front cinch and on each of the rear side straps is a small metal ring. The llama is told to kush and small diameter ropes are tied from each of the metal loops to the seat fasteners located on the floor along the sides of the plane. The two ropes on each side of the llama prevent him from raising his hips or shoulders. The neck collar is adjusted so that the small metal clip at the bottom can be directly fasten to one of the seat attachments in the center of the floor. The animal is then completely immobilized in his normal, comfortable lying position. Even llamas new to flying will try only two or three times to get up before realizing that the seat belt has them securely bound to the floor of the plane.

Although the Cessna 206 will usually haul only one big llama, there was plenty of room in back of the llama to place saddles and panniers. The Cessna 206 has the advantage of a rear side cargo door which permits easy access to the rear compartment. Larger planes, such as the Cessna 207 or 208, that land on wilderness air strips can haul a couple of llamas at one time; some other planes could haul three llamas. In order for llamas to board a plane, the wings must be above the door and not below the door.

Mike's experienced llama entered the plane with little difficulty; however, both of mine had to be encouraged. The first one decided that jumping up into the plane with the wing over his head just wasn't the thing to do. A couple of nearby wooden loading pallets quickly remedied this problem by reducing the height of the jump. I'm sure that they would load much easier the next time.

Although the llamas tolerated the flight, I can't say that they enjoyed it. It was either the bumpy ride or the wonderful scene of the Big Horn Craigs all topped with snow that took my breath away; the turbulence over the near 10,000-feet peaks caused the ride to be rougher than what I enjoy. During the rough air, the llamas responded favorably to a familiar, calm voice. During a few of these times, Bandit rested his big head on my hand as if to say that he was placing himself in my hands - hoping that I knew what we were doing. Perhaps this trust was similar to what an experienced packing llama develops in his owner, following almost anywhere.

The Frank Church River of No Return Wilderness covers 2.3 million acres in central Idaho. It is the largest continuous wilderness in the lower 48 states. As we flew along, Mike explained that in the legislation that established this wilderness was a provision to leave the existing landing strips in place, permitting use of planes on the strips and jet boats on the Salmon River. In some areas, as in the Middle Fork area, there is a landing strip every seven or so miles. These strips are usually snow free and useable from mid May to the end of November. The Salmon River and its tributaries drain this hugh wilderness and in the process have cut the second and third deepest canyons in North America. The elevation varies from 2000 feet at the River to 10,082 at the highest peak.

Our goal was the 5500-foot landing strip at Chamberlain Basin, a well known drop off point for hunters in the fall. It is located some 25 miles into the wilderness and 80 miles from Salmon. Elk were grazing on several parts of the air strip but none were in our landing path. Back country landing strips are favorite places for elk, and occasionally the pilot must pass low over the runway to chase them from the landing area.

After each smooth landing on the grass covered strip, a llama hopped out and immediately started eating the green grass as if nothing unusual had occurred.

By the time we had tied down the plane, erected the tent and cooked supper, it was dark, and the cool breezes suddenly turned cold. We dropped off to sleep with our heads covered and the winds whispering to us through the tree tops. We woke the next morning to an overcast sky, heavy frost and solid ice in the coffee pot.

After a quick hot breakfast, we were anxious to be walking in order to get warm. However, our progress for the next three hours was slow because we had to stop at every opening or meadow and look at elk with our binoculars. We must have seen over 500 elk feeding in the early morning shadows.

Himalaya followed me up the steep plank into the jetboat on a loose lead like he had been doing it all his life.

The steep trail down Chamberlain Creek proved more interesting than we had anticipated. A forest fire had followed the creek for a number of miles the previous summer. In some areas, the downed timber prevented us from using the trail. In one area, we waded in hip deep water because of downed trees and high water.

The downed trees were often a challenge. We could step or jump over many of them. It was impressive to see the llamas jump these high obstacles. Some were too big to cross or were suspended in the air so we had to find a way around. I gained a new appreciation for llamas and their abilities to traverse steep terrain. Some of the ridges we climbed over to get around fallen trees were so steep that we traversed them in a lazy S pattern. Once we had to go below and around a fallen tree above a precarious drop off overlooking Chamberlain Creek. The fire had left a foot of loose soil around the trunk of the overturned tree. Because of a rock outcrop, we had to climb straight up the steep slope, and I used my hands in order to make it. The llamas were unbelievable as they lunged with all four legs and humped their back as they continually leaped upward to the trail.

The trail follows Chamberlain Creek as it drops rapidly and continually downward toward the Salmon River making only a few spots flat enough to hold a tent. After spending another night and day on the trail, we completed, in complete isolation from other humans and problems associated with civilization, the 22-mile trip down Chamberlain Creek. All too soon we were standing in front of a sea of boulders lining the Salmon River.

We walked our llamas through the boulders and unpacked them on the rocks near a relatively calm spot of water. Then we waited for our prearranged ride by commercial jet boat. Finally, from around the bend, we saw the jet boat rapidly coming downstream. The boat came along side, and we tied it to the boulders.

A one-inch thick four by eight-foot plywood sheet was extended over the high rail in the rear of the boat to the rocky surface. Three 2 by 4s had been nailed across the eight-foot length to give the llamas some traction. Although the slope

Mike Tupper enjoys the scenery as we jetboat up the Salmon River. The llamas kush calmly in the boat.

of the board was 45 degrees and the llamas had never climbed anything like it, they followed us up the plank on a loose lead, just like they had been doing it all their lives. We quickly jumped them from the paneling over the motor down into the bottom of the boat. We quickly kushed the llamas, as the boat pilot touched the throttle and the big, nosy 460-cubic inch ford engine jumped to life. We were soon followed by a 30-foot rooster tail of water as we began our trip upstream to a road.

When the big jet boat climbed up the numerous rough rapids, the big loud jets changed to a equally loud wine as air occasionally replaced water in the jets. The loud noise didn't phase the calm llamas and only twice did one of them begin to get up, but quickly laid down again on command. The 20-mile trip through the steep picturesque canyon of the Salmon River took only about one hour. When the pilot opened the window on the front of the boat, we and the llamas climbed through the small opening onto the metal covered bow.

The scenery had been spectacular, the air pure, the flowers beautiful and the llamas had performed almost flawlessly even on this initial trek of the year. The llamas, who had so easily flown and jet boated in and out of this magnificent wilderness, had added considerable to our enjoyment and in their silent and obedient manner had contributed and had not taken away from the reasons people visit wilderness areas. As we turned to watch the river flow back into the wilderness, we briefly paused and relived in our mind's eye the wonderful experiences we enjoyed in the solitude of the wilderness. Although we'll surely make additional flying/boating llama treks into these wildernesses, the first one will hold an unrepeatable place in our mind. For us, the mighty Salmon River really was the River of No Return.

CHAPTER 29
CHOOSING A LLAMA VETERINARIAN

Whether you own llamas as pets, packers, investments or all of these, the health of your herd is of utmost importance. One of the points that we llama lovers like to stress is that llamas are healthy animals who seldom get sick or need medical attention. Although this is true, anyone who has llamas will eventually need some type of veterinary services. Perhaps you have already found a veterinarian with whom you have confidence to assist you in providing health care for your llamas. However, many new owners have yet to make a selection. In more remote areas, you may not have access to more than one or two veterinarians. Several people I know have several vets in their area, but only one has agreed to consider treating llamas. Luckily, this particular vet is interested in llamas and is well liked by his llama clients. If you are fortunate, you'll have several vets from which to choose. So how do you go about choosing a vet for llamas?

All veterinarians have Doctor of Veterinarian Medicine degrees and have passed and sufficiently mastered some basic skills and procedures through courses and internships to gain their degree. In addition most have been required to have certain skills, courses and a Doctor of Veterinary Medicine degree to obtain a license in the state in which they practice. These are basic requirements for any veterinarian.

The vets in your area likely graduated from different universities. However, place of study may not be an adequate reason to choose a veterinarian. There are a number of good veterinarian schools in the United States and Canada. Most of us are not capable of judging which school is the best. In fact, even so called experts in this area would not agree as to which schools produce the best veterinarians. All of the veterinarian colleges graduate DVMs which are great and some which are not as good.

A vet from a school that has an active llama teaching and research program should be carefully considered. However, they may not have encountered llamas in their internships unless they graduated fairly recently. Other criteria may be more important.

Although basic education and training may be similar, their knowledge, experience and skill can be different. Some small animal practices will consider treating llamas. However, a vet who has at least some part of his practice devoted to the treatment of larger animals would be more skillful in treating llamas. Setting a broken leg, treatment for cuts, external and internal parasites, dystocia, castration, ruminant digestive disorders and many other treatments are similar among various large animals. Therefore, a vet in a large animal practice already has some of the knowledge and experience necessary even if he has never seen a llama.

Asking friends and acquaintances about veterinarians may provide some insight into their skill. However, be careful about this type of advice since this may not be a reliable indicator. People can be very emotional about animals and this might impair their judgement. Their negative or positive opinion may be based upon an extremely small sampling of the capabilities of the veterinarian. If there are llama owners and breeders in your area, consult them about the vet they use. If it can be done diplomatically, consulting with another veterinarian about vets in your area may also be helpful. If you are not acquainted with anyone using your prospective vet, ask him for names and telephone numbers of current clients who use his services regularly so you can call them as references. A personal discussion or interview with a vet can reveal much about his or her attitude, personality, approach to their practice, their potential interest in treating llamas and their experience. If you choose this particular route, make the interview short. The larger llama breeders who require services of a vet use these techniques of references and interviews in finding the right vet and you should also.

One of the key factors is personality. If you like your veterinarian as a person, you will likely listen and communicate freely. Your veterinarian may be the most knowledgeable person in the world about llamas, but if you can barely tolerate being in his presence, you should certainly consider changing vets. Life is too short and llamas are too wonderful not to enjoy all aspects of llama ranching. I fully realize that llama owners in general are some of the nicest people. However, all of us have our own unique personalities, and the vet/owner relationship also depends upon your own personality. If your veterinarian appreciates you as a person, the relationship will more likely promote honest communication.

Your veterinarian, especially if he is new to the llama practice, should be willing to seek advice on a difficult diagnosis. In my opinion, this is one of the key characteristics in your search for a veterinarian. He may not know who to ask. This is where your knowledge should be helpful. Through your reading and attending various local, regional and national conferences, you should be armed with some suggestions as to whom to contact. If you are new to the llama world, then call and ask the breeder where you purchased your llamas. Most serious breeders welcome and expect to answer questions. A confident vet can do wonders in relieving our worries about a

CHOOSING A LLAMA VETERINARIAN

llama. However, if your veterinarian seems to be an instant expert and pretends to know everything there is to know about llama health, perhaps you should reconsider.

If you have a larger llama herd or if there are several other llama owners as clients, your veterinarian should be willing to attend some of the medical conferences and seminars on llama medicine. You should provide him with information on these type of events.

Convenience is an important consideration. If you need the service of your veterinarian in an emergency and his office is several hours away, this could lead to unpleasant consequences. Being a few minutes from your vet's office can be a real time saver for routine visits. If you live on an isolated ranch in eastern Montana or southern New Mexico, you may be accustomed to the inconvenience factor. However, for many of us, having to wait quite awhile for a veterinarian to arrive may have us climbing the wall during an emergency.

Perhaps you will be as lucky as Bill Garrett of Madison, Indiana in finding a llama veterinarian. When Bill acquired llamas, a young vet who lives nearby and who had just joined a local vet in his practice, showed up at his ranch to see and talk about llamas. Subsequently, he has showed a real interest in all aspects of llama health. Being genuinely interested in learning about llamas is one of the real keys for making good llama veterinarians.

Many of us llama breeders have full-time occupations away from our llama operations. Feeding, llama care and training are generally handled after regular hours and on the weekends. These are the times most of us notice health problems. So you may need to know if a vet has Saturday morning office hours and what are his normal office hours during the week. You may not enjoy having to take annual leave from your job for a routine office visit. On the other hand, don't expect your vet to periodically meet you after hours to handle routine medical problems. Another crucial item is the type of emergency after hour service that is available. Is there a difference between what is stated and what is actually practiced? If you know of someone who has used this emergency service, you can ask about their experiences.

A veterinary clinic with two or three veterinarians does have some advantages. This is particularly true with emergencies. Generally, a multi-vet office has someone on call at all times. Taking turns with emergency calls permits the vets to have time with other activities and helps make them more responsive to emergency calls. Another advantage is one of the vets is generally at the clinic during office hours. They can easily cover for each other when vacations, sickness and other situations prevent them from coming to the clinic. If the veterinarians are mature individuals, they can also benefit you as a llama owner by consulting with one another over a particular llama problem.

We use a clinic that has two young lady veterinarians. Both treat our llamas so we feel comfortable with both. Although I'm sure neither would prefer to come out on a cold subzero night to treat a sick llama, both are willing and cheerful about doing it, even when eight-months pregnant. One night when one of the vets was tied up at the local horse racing track, we needed some emergency treatment for one of our females. The other vet was home with her week-old child. Since she did not expect to be called out so soon after the birth of her child, she did not have an available sitter and her husband was away from home. My wife went to her home and sat with her two sleeping young children, while she met me at the clinic.

CHOOSING A LLAMA VETERINARIAN

When we received her bill for her services, almost half of the charge was subtracted for baby sitting. Service with a smile, you bet!

Although cost is a factor for most people in choosing a veterinarian, it should be considered along with the other selection processes. No matter what our economic condition, we expect charges that are fair and equitable. Most veterinarians realize this and they or their receptionist will be glad to give you a general idea of the cost for their basic office and ranch calls. If cost is important to you, don't be embarrassed about asking. They would much prefer that you understand their price structure before you begin to use them rather than complain about it after treating your animals.

Another factor closely coupled with cost is whether or not your potential veterinarian normally practices out of an office or clinic. Some veterinarians are finding they can be of better service to their large animal clients and their practice can be more profitable if they operate (no pun intended) out of a well- stocked truck. This type of practice is probably wider spread in the west where clients are located in more remote areas. Generally, when an operation involving specialized equipment is needed, they make arrangements with other veterinarians to use their facilities. This type of service may be to your liking if you prefer the vet to come to your ranch for everything. This eliminates the need to catch and load your animal and take it to his office. This may also eliminate the need for you to own a truck and/or trailer thus reducing your equipment cost. However, before committing yourself to this seemingly delightful way to see your vet, consider cost and a potential delay in contacting your vet in an emergency. I can bring three or four of my packing males in to the clinic for a health certificate for travel into an adjacent state and perhaps be charged $10 (assuming no blood or TB testing is required). On the other hand, this would cost a friend of mine $32, the standard farm service call for a visit from his mobile vet. If your vet travels all over the country for routine calls, he may not be able to immediately come to your place to assist in a difficult birth or some other emergency. Vets with clinics also make ranch calls as part of their practice. You'll need to evaluate which service you prefer based upon your own particular situation.

You may want to do some of the simple maintenance procedures yourself. You may need for your vet to demonstrate the proper techniques and precautions. On the other hand, you may not be capable or have the time or desire to do these tasks. Although it would probably be rare, a vet with a large practice may not find the time or else charges too much for the service of doing routine maintenance on your llama. Be sure the vet you choose is willing to offer advice on these matters.

Don't wait until you have a life threatening situation to become acquainted or to at least have chosen a vet. Try to establish some sort of relation with your veterinarian as soon as you acquire llamas if not before.

Once you have chosen your veterinarian, there are several things you can do to keep good relations. One is to promptly pay him for his services. Most of us would probably be reluctant to sell another llama on credit to someone who took an unreasonable amount of time to pay for a previous purchase. At least, we would be more inclined to sell our best stock to someone else. Likewise, I can't imagine them giving you their best service and highest priorities if you seldom pay your bill on time. Nearly all of us have had money difficulties, and your veterinarian will probably understand if you need to pay your bill by installments, particularly for some expensive treatment. Be honest with your veterinarian or his representative and let him know how you are planning to pay him.

Another way to keep a good relationship with your vet is to notify him before a problem becomes severe. It is easier to treat and correct a problem, and usually cheaper, before it becomes a major problem.

Also remember that veterinarians are not perfect; they do not know all the answers about the treatment of llamas. Even the best llama experts do not always agree or make the proper diagnosis because llama medicine is somewhat new. Sure your vet will make some mistakes and will occasionally make the wrong diagnosis. Be reasonable in what you expect. If the problem is not cured the first time, give him a chance to try something else. Learning along with your vet can be a rewarding and interesting experience.

Be reasonable in what information and diagnosis to expect via the telephone. I know of a person whose animal died, and he claims the vet did not know how to threat it. Although he discussed the situation on the telephone several times, he never actually brought the animal in for the veterinarian to see. The vet can do a better job of diagnosing your animal's problem if he can actually see it. We may not describe the animal's problem correctly or notice other accompanying symptoms. Many conditions look similar and the owner can misinterpret the signs and cause the vet to think that the problem is better or worse than it really is.

Choosing the right vet to treat your llamas is an important tool in your management program. The selection process shouldn't be by chance alone. Management of your herd can be easier and more fun with the right person assisting you in maintaining healthy and productive llamas.

CHAPTER 30

LLAMA HEALTH AND CARE

Although llamas seldom need any kind of health care, there are some routine things you should do for your llamas.

If you have been fortunate to hear some of the top llama veterinarians speak on llama health, you may have noticed that at least some of them seldom recommend the type of vaccinations to give your llamas. They may have indicated that animals' needs for vaccination differs between regions, and you should consult your veterinarian for specific advice. I'll tell you with what and when I vaccinate, in order to demonstrate how simple the vaccinations and procedures can be, but I <u>don't</u> recommend that you do it this way. I do recommend that you consult your veterinarian before vaccinating your llamas!

If you have attended any of the llama conferences, you may recall that Dr. Murray Fowler informed his listeners that the effects of vaccinations on llamas have not been studied. There is no research evidence that the vaccinations given to your llamas are effective. That doesn't necessarily mean that the vaccines aren't giving some protection to your llamas. I know the vaccines make my cattle-raising neighbor feel a lot better about having llamas next to him.

We vaccinate each llama in the spring with either seven- or eight-way. The vaccine includes **Clostridium chauvoei** (blackleg), **C. septicum** (malignant edema, **C. haemolyticum** (bacillary hemoglobinuria/red water), **C. novi** (black disease), **C. sordellii & perfringens** (type C & D, enterotoxemia). We also vaccinate for tetanus. We use an Ivermectin vaccine to worm our llamas in the fall. It is common livestock procedure to vary the kinds and types, oral and vaccine, of worming medicine. This prevents a parasite from building a resistance to a particular type of drug.

The number of times you need to worm your animals usually depends upon your location and climate. You should consult your local veterinarian for this information. Usually the wetter the climate, the more times you'll need to worm. He can also help test your llamas for worms.

Your veterinarian can show you the proper method for vaccinating. Some veterinarians recommend changing both syringe and needle after each injection. Although we always change needles, we often use the same syringe with the same bottle of drugs for several llamas. You should consult your veterinarian for his recommendation. Although any large muscle mass is suitable for giving intermuscular injections, the front shoulder and rear legs are often chosen.

When a male is about two and a half-years old, you may find that his companions occasionally have blood marks on their ears, back of the neck and on the front and rear legs. When this happens a few times, you know that you need to do some dental work on your llama. Your llama may or may not have all six fighting teeth erupted, and the fighting teeth are small. Although they may be too small to inflict serious harm, they are extremely sharp at this age and often need to be cut. Although this initial cut will get rid of the sharp points, you may need to repeat the procedure at a later date. The fighting teeth will continue to grow for several years after they have been cut, but they will have a flat, smooth surface and can cause little harm to other llamas.

HEALTH AND CARE 240

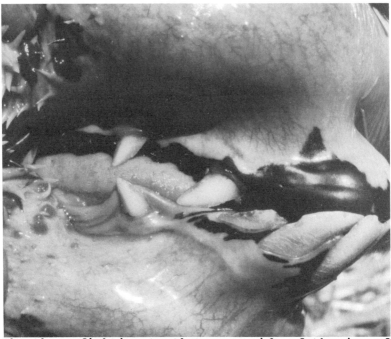

The three fighting teeth on one side of the jaw of a young male.

The purpose of the fighting teeth is, of course, for fighting. Their quanaco ancestors used them to defend territories and to protect their harem. In llamas they serve the same purpose. Fighting usually consists of rearing and hitting with the chest and necks, biting of the front and rear legs and biting in the back of the neck. Most of the fighting takes place with the adolescence studs, which usually does little harm, and mature studs.

We usually need to trim our llamas' toe nails twice each year. We normally have lots of snow, and snow doesn't do much for keeping toe nails worn down. So in the spring, we trim all of their toe nails and give vaccinations at the same time. We also trim nails again in the fall and worm them. If you have rocky soil and little snow, you may not have to trim toe nails quite so often. Males that are packed often in the summer and fall usually do not need any trimming until spring. This regime forces you to handle all your llamas at least twice per year.

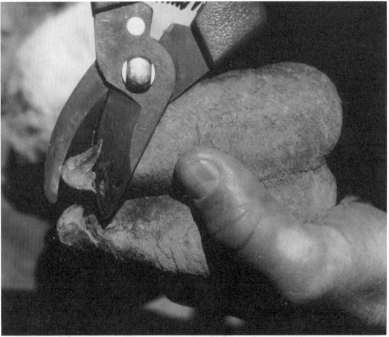

Brush clippers do a good job of cutting the toe nails.

If you trim toe nails when the soil and vegetation are still moist, the nails will be tender and easier to clip. The equipment for nail clipping is inexpensive. A $6 to $8 brush-hand trimmer is all that is necessary. We've trimmed over a 100 sets of toe nails with one tool without it becoming too dull to use. The object is to trim the nails back to the fleshy part of the pad. You can use the point of the clippers to clean out the mud and dirt under the point of the nail so you can tell how close to trim the nail. There are two toes and nails on each foot. If the nails aren't too long, three basic cuts may be all that is necessary. One cut along each side of the nail and one across the front of the nail may do the job. I often make several cuts on each nail, thus, making sure that the fleshy part of the pad is not injured. We have seen nails that were neglected and grown long and twisted to one side. When this happens, the nails on top of the toe also twist to one side and frequent trimming may be necessary to correct the problem.

Most of our llamas stand reasonably still as we pick up their feet and trim their nails.

Many llama breeders use various types of llama restraints. We have used at least three different types of restrainers for vaccinations, cutting fighting teeth and for trimming feet. It is a common piece of llama equipment, and many breeders have them. If you have a reasonable animal, they work fine.

Some of the restrainers have straps or cinches that go under the animal to keep him or her from lying down and another one or two go over the back to keep the llamas from rearing up in the chute or restrainer. This same type has a head restraint that goes under the chin of the animal and is adjustable for height. Another general type of restrainer has two poles in a v-shape. The llamas's head is positioned in the notch of the v and the poles are closed in on the neck. Two leads pull the animal forward in the restrainer.

If you have an excitable animal or one with little or no training, your llamas can be seriously hurt in a restrainer. For the past two years, we have used our restrainer, not as a restrainer, but as a holder for our platform scales. The restrainer's sides help keep the llama on the scales.

Unless you have handled llamas extensively and have seen restrainers used, I suggest you be cautious about using one, and perhaps postpone your purchase of one. Now don't get me wrong, some restrainers may work great and be completely safe. However, I think for most llama owners, they aren't really necessary.

One of the most effective restrainers I have seen is a gate and a wooden fence. The llama is placed against the wooden fence. The metal or wooden gate is then pushed against the animal and shots and other health maintenance is done. I think it is one of the cheaper and safest restrainers I have seen. Some owners have a wooden head restrainer consisting of two parallel pieces of wood that are permanently bolted at the bottom and the top opens for entry and closes tightly to restrain the head. Often owners don't use a head restraint.

If restrain is necessary, we use a restraining device that is commonly available on every llama owners place - a fence. Cherrie and I have used a fence as a restrainer when we vaccinate, trim toe nails and cut out fighting teeth. We have cut out fighting teeth for different llama owners throughout southeastern Idaho and have always found a suitable fence to help restrain the animal. It works great, it's safe, it's quick and we can have a llama's fighting teeth cut out and have him loose in the pasture before we could get a llama strapped into our restrainer. Almost any wooden fence will work to help restrain the animal. The preferred place on the fence is against a corner. The corner helps prevent the animal from moving forward. Some owners prefer to use the inner side of their stock trailer to help restrain an animal. We have used our trailer for this and it works fine.

If your animals have received the proper basic training and have had their toe nails trimmed a few times, it may not be necessary to restrain them at all. Recently, Cherrie and I trimmed the toe nails and gave an ivermectin injection to 32 females and three shots to 18 babies in less than two and half hours. On the adult females, Cherrie looped a lead around a post and held them near the post. I lifted all their legs and trimmed each nail without restraint or difficulty.

We put a single wrap of the lead around a post so that the lead can be instantly dropped if the animal gets into trouble, thereby freeing the animal's head. The llama is then pushed against the wooden fence, pulling his head snugly against the post for tooth trimming. I stand at the male's shoulders pushing the animal against the fence while holding the lead. I then put a arm around the llama's head with my hand across the nose and lift the upper lip. With my other hand, the lead is held across the neck helping to hold the animal steady. This hand is on the bottom of the lower jaw and opens the lower lip. This ensures that the lips will not contact the rapidly moving wire saw, and exposes the teeth for easy cutting. A helper stands and pushes the llamas's middle and rear against the fence. Holding the llama in this position is fairly easy and does not require much of an effort. Cherrie then uses a surgical wire (about 12 inches long) with handles on each end to cut the six fighting teeth (two on the top jaw and one on the bottom, on each side of the mouth) level with the gums. Once she has the wire in position on the tooth, she pulls hard and very rapidly moves the wire across the tooth at the gum line. On the bottom fighting teeth the cutting wire handles are held level or slightly below the level of the jaw and on the upper fighting teeth, the handles are held level or slightly above the level of the jaw. This assures that the wire doesn't slip off the tooth and that the tooth is cut level with the gum. Usually in less than a minute, she has the three teeth on each side cut. Then the animal is reversed and the procedure repeated. No cooling water or anything else is required. When Cherrie cuts them off, there is usually no bleeding. However, a few drops of blood on the gum near the tooth may form, but this is not a serious problem and will stop bleeding quickly. The complete job usually takes only three or four minutes from start to finish. In most cases, the llama is not unduly disturbed and acts completely normal when the procedure is completed.

Be sure that you examine the tooth to see if you have left a sharp edge on it. You may need to file it smooth with a small metal file or fabric coated sand paper. After you have cut a few, it is rare to have a sharp edge.

When we trim toe nails and give vaccinations, two people are usually all that is necessary. However, I have vaccinated a number of our tamer geldings and studs by myself. Again we use the wooden fence. The lead is wrapped once around a post, again so that just dropping the lead releases the animal. The lead is then held on the outside of the neck to help restrain the animal, and the llama is pushed against the fence and the shots administered.

The same general procedure is used for nail trimming, except you may find that plywood against the fence helps prevent the animal from climbing the wooden rails or from sticking their legs through the fence.

In a previous chapter, we have indicated that both studs and geldings can be used effectively in packing. However, no matter how happy you are with the performance of a mixed stud and gelding pack string, you or other members of your family may not appreciate the behavior of a corral or pasture with mature studs or a mixture of studs and geldings. I know of breeders who keep their studs in the same pasture without too much difficulty. Whether or not you have females, stud llamas are going to fight more than geldings. The amount of fighting, depends upon individual animals and the size and shape of your pasture or corrals. Some have made the point that housing your studs together, keeps them in good shape because of the fighting and chasing. This is no doubt true, but I personally don't enjoy seeing adult llamas chasing and fighting. Therefore, I pen my adult studs separately. By the time they are three- or four-years old, you may need to make a decision about whether or not to geld them, pasture them together or to pen them separately. If you don't plan to use them for breeding, this is a good age to geld them. My geldings and younger studs run together in a pasture next to a pasture of females without serious fighting. A group of geldings are normally content with each other and do very little fighting.

Fighting and hollering llamas do not make good neighbors nor do they have a positive effect on public relations for llamas. Your veterinarian can quickly geld your studs. Our vets charge $45 per animal, but the charge will vary between vets and location. Depending upon the individual, it often takes one to two months for the gelded mature animal to quietly and non-aggressively take his place with the geldings and younger studs.

From time to time, your llamas will have small cuts and scratches that you should be able to treat effectively by spraying with an antibiotic.

If you own only a few llamas, you don't necessarily need to be able to cut out fighting teeth. If there is a breeder in your area, they will often do this task for you. It may be inexpensive for your veterinarian to give vaccinations for only a few llamas. You should also talk to your veterinarian if you plan to administer your own shots. Although rare, under the right conditions, a llama can go into shock if the vaccine is placed in a blood vessel. Your vet can supply you with another injection that will counteract the shock.

By some simple procedures, you can keep your llamas in good health. I encourage you to depend upon your veterinarian for advice and assistance.

CHAPTER 31

EMERGENCY AID ON THE TRAIL

Llamas are tranquil, calm pack animals so they are less likely to hurt themselves than other pack stock. You'll likely go many years with a pack string before you need any first aid for your llamas while you are packing. We have never had any problems with our llamas requiring first aid treatment while out packing, hunting or fishing. Inquiries to other commercial and non-commercial llama packers receive similar responses. Some have had one or two minor injuries or sores, but most report that they have never had a problem. However, you should be prepared to handle a few emergencies. Should a problem develop, you should be able to provide enough emergency aid for your llama to get him back to civilization so that you can take him to a veterinarian, if necessary.

I recommend that you take a human first aid or emergency aid course. Admittedly, you can't necessarily treat llamas like people, but the basic information on cuts and sprains will be of value to you as you treat a llama problem.

We have had two or three cuts on llamas, but most of them occurred when we rented a place for some of our llamas, and it came with built-in hazards. Most abrasions, minor cuts or wounds can be treated with a topical medication. Our veterinarian recommended a spray which is easy to apply. Usually these types of wounds heal quickly and aren't much of a concern.

Deeper cuts especially on the legs may need more attention. If bleeding does not stop within a few minutes or if severe bleeding occurs, you must help stop the bleeding. This can usually be done by placing a clean cloth or gauze on the wound and applying pressure which decreases bleeding and allows the blood to coagulate. If the wound is dirty, try to clean it with clean water or a combination of clean water and mild soap or a commercial product such as Betadine pads or swabs.

If the cut is on a leg or foot, the cut should be covered. Place a piece of gauze or cloth, if you do not have gauze, on the wound and wrap with a self adhesive wrap which is commonly available. A layer of duct tape can then be placed over the wrap. This will keep the wound clean until you can get back home or to a veterinarian. You should wrap the cut tight enough to keep the gauze in place but not so tight as to hinder blood circulation.

Depending upon distance, original weight carried when injured and severity of the cut, you'll need to make a common sense decision on reducing the load and distance to be traveled. for this particular llama. If the injury is severe with contamination, your veterinarian may want to give systemic antibiotics to the llama when you have returned. Any cut where the skin gaps apart may need sutures.

A cut or puncture wound on the pad of one of the toes can be a big problem. Previously, when we rented, one of our llamas ran across a broken beer bottle and cut one pad all the way across. Treatment consisted of cleaning the wound with clean water and mild soap, applying a topical medication, wrapping with gauge and a self adhesive wrap. The bandage was changed periodically, and the animal was placed in an enclosure to minimize movement. Usually with a severe cut of this type, your veterinarian will want to give systemic antibiotics. This stud was not available for packing for the rest of the summer. Eventually the pad was replaced with a complete new one. If a severe cut on the pad occurs while packing, you'll need to slowly lead the unpacked llama back to the trail head after you apply treatment. A boot made from rubber, leather, canvas or carpet pad might be helpful in getting the llama home. A boot can usually be made out of something you have along on the trek. On any cut or wound, you should make sure that your llama has up to date tetanus vaccinations.

Usually if your llama has a cut that needs to be sutured, you can return to the trail head immediately and take your llama to the veterinarian. Although it would be very rare, a suture needle and suture material might be necessary to use in a severe cut to close off a blood vessel.

Minor bruises usually do not affect the packing of a llama. For some major bruises, especially when you can see some swelling, you'll likely need to lightened the load. Any bruises that cause fluid to build-up should be tended by a veterinarian as soon as possible. Occasionally llamas sprain a ligament or obtain a severe bruise that will keep them from packing for several months. We had a gelding injured on ice one winter, and he limped for several weeks. If a minor fracture occurs, you may need to splint a leg and slowly walk the animal back to the trail head. Wrapping the leg with carpet pad or some other material prior to splinting might be helpful.

Although saddle sores are rare for the weekend or vacation packer, they can occur in commercial or long-term packing. They can develop into serious problems and should be dealt with as soon as possible. Extra padding, for example with foam, usually helps saddle sores. Severe sores, that is raw or red or pink in color, should be treated with topical medicine. If possible give the llama time off to heal. Sometimes straps or cinches can be tied together or in some other way tied to other parts of the saddle to force the cinches away from a cinch sore.

Although heat stress or hyperthermia is more common in hot and humid climates, it can occur in other situations. If you plan to pack in hot climates or in the middle of the day during unusually hot spells anywhere, you may want to consider shearing some of the wool. Climbing and very heavy loads can produce hyperthermia in hot weather. The work and humidity can cause the core temperature to rise above normal.

Panting, irrational or uncoordinated walking are often signs of heat stress. If you suspect heat stress, you should stop immediately, remove packs and put your llama in a shade. If water is nearby, splash water on the llama, particularly on the stomach and other less-wooly areas. The insides of the back legs have two large blood vessels, and my vet indicates that, at least in horses, this is one of the recommended places to put water. This can be a life-threatening situation. We have never experienced heat stress as most of our packing has been in the intermountain west where humidity is usually low, and we rest our animals more often in warm weather. However, it can happen anywhere and can result in death.

Heat stress can also occur in a hot or poorly ventilated trailer. If you suspect the temperature is too warm in your trailer, wet down the straw or other bedding and the llamas periodically.

We do not carry a thermometer with us when we pack, but if you do, you can check the temperature of your llamas if you suspect hyperthermia. Normal temperatures should be around 99 to 102 degrees fahrenheit. Elevated temperatures above these values are usually a sign of heat stress.

Puncture wounds are normally cleaned and a topical medication applied, and if not severe, left uncovered. A puncture wound to the foot will likely require slowing the pace and lightening the load.

Rattlesnakes or other poisonous snake bites can be fatal to llamas, especially bites on or near the nose area. Since llamas are so curious, they might not avoid rattlesnakes. If a bite on the face occurs or if you see the nose swelling, suspect a snake bite. Watch the nose area carefully, and if it begins to swell, you'll need to insert tubes in the nostrils so that the swelling will not suffocate your llama. You can have tubes available for this purpose or you can use the tubes on your water purification kit.

Flies and mosquitoes can be especially bothersome to packing llamas. Some type of livestock insect repellant can be placed on the llama. Generally it can be sprayed on the legs, rear, chest and other less wooly parts of the body. However, for the face, put some of the liquid on a cloth and rub it on the llama or use a roll-on type repellant. Never spray insect repellant toward the eyes!

Ticks can be a severe problem in some areas. They are often the most prominent in the spring. I don't often pack in March and April. However, I have administered ivermectin shots before trekking in the spring into areas with heavy tick infestation. This provides some temporary protection for llamas against external parasites. It kept the ticks off my llamas, while a companion's llama, who did not receive ivermectin, had several attached ticks.

I have only administered the shot once to individual llamas for spring packing in tick country. If you plan to pack several times in the spring, I'd suggest that you contact your vet for possible tick preventative measures.

Some ticks have a toxin that can paralyze your llama. If your animal becomes uncoordinated and begins to be paralyzed and heat stress is not a factor, you should immediately suspect tick paralysis. It is important that you find the tick(s) and remove them immediately. Tick paralysis can be fatal! Usually the ticks are on the less-wooly parts of the body, such as underneath, ears, back of hind legs, anal area and the leg-body connecting areas. If you find all the ticks, recovery can be quite rapid. However, you may have to allow the llama a day or so to recover before you head back to the trail head. In a pasture situation, tick paralysis can be treated by both tick removal and a subcutaneous (under the skin) injection of ivermectin which is effective in ridding the llama of any ticks you may have missed. One year, we took three females to Montana for breeding. One of the females became paralysed. Although a tick was not found, a subcutaneous shot of ivermectin had the female up on her feet very quickly and back to normal in only a few days.

Any debris found in the eye should be removed. Usually a corner of a clean cloth or towel can be used to lift out a spec or two of material from the eye. Moistening the cloth might help. Once a female of ours needed some material cleaned from her eye. Her eye was swollen and obviously painful to open. We hadn't owned the llama long, and she was still quite wild. However, she calmly let us remove the two pieces of debris from that eye. However, she would almost throw a fit if we even approached the other eye. Obviously she wanted and allowed the attention on the injured eye. A syringe can be used to flush out particles from the eye. See your vet for advise on some type of ointment for the eye to carry along with you should an eye injury occur.

Some items handy for your llama first aid kit include:
- vet wrap or coban wrap (self adhesive wrap)
- duct tape
- antimicrobial ointment such as Furacin or Betadine (to treat wounds and cuts) or 7% iodine or other topical medication
- Butazolidin paste or powder (analgesic-used only as necessary to reduce intense pain; the powder is easily mixed with water and inserted in the back of the mouth with a syringe)
- leather or inner tube rubber
- Ophthalmic ointment (treatment of minor eye injuries)
- Suture needles and suture material (purchased through your vet or monofilament fishing line)

This is meant to be a minimum list. Other items can be found in your human first aid kit. You can visit with your veterinarian for other suggestions. This list was modified from one provided by Dr. LaRue Johnson at a 1989 llama conference.

CHAPTER 32

CATS AND LLAMAS

Cats can provide several beneficial services to the llama owner. When we rented a place to keep our llamas, we had mouse problems. We kept our sacked feed in an old chest freezer. However, even with the freezer latched except when feeding, we occasionally found mice in it. We frequently saw mice in the hay barn. We placed several containers of mouse poison in various places in the barn. We couldn't tell that the poisons were effective in reducing the number of mice. Perhaps too much other food was available to repeatedly attract the mice to the poisoned baits.

A small white female cat came to stay at the barn. We fed the cat regularly, and within a few weeks all traces of mice were eliminated. She hunted the fields after she had cleaned mice out of the barn.

She was the nicest cat! She had a litter of kittens each year. When the kittens began to move around, she would take them all to a neighbor (her original home) who had lots of cats and then would leave them and return to us.

Not long after we moved onto our new farm, a kitten was dropped off by someone. This little cat, now mature, has continued in the fine farm cat tradition of keeping the mice out of our hay shed and barn. Recently we spent five hours with friends clipping toe nails and giving ivermectin shots to our llamas. We used 115 needles during the day so we were busy most of the time. However, one of the ladies noticed one of our cats going into the pasture six times. Each time she returned, she had a different mouse. The next day, we

CATS AND LLAMAS 256

Llamas enjoy cats and cats serve a useful purpose around the llama farm.

were at our llama farm about 20 minutes. The same cat came out of the pasture with another mouse. So cats can be extremely effective predators, and therefore, valuable assets to your llama operation. Unfortunately, under some conditions, studies have documented that urban cats harvest a surprisingly large number of birds.

Cats also provide a lot of entertainment for both the llamas and Cherrie and I. Since llamas are very curious creatures, they often follow our cat around in the pasture; they watch her climb around in the corrals. Baby llamas in particular, appear to enjoy trotting behind the cat. We enjoy the interaction of the cat and the llamas.

Our cat follows along with us when we walk into the pastures each day. We enjoy her company. Last summer, I was walking with the cat in the pasture. Every few steps, the cat would run in front of me, fall to the ground and roll over on her back. After repeating this several times, I finally figured out what she was up to. She rolled over because she enjoys the llamas nuzzling her tummy when she rolls to her back in front of a llama. She expected that I would do the same thing.

We feed our cat dry cat food. Our small cat does not eat very much, and she supplements her commercial cat food with mice. Cats should be vaccinated each year and wormed occasionally. Your vet can provide these services or vaccinations and worming medication can be purchased at your local livestock supply company. Then you can very economically give the vaccinations. Since our cats provide a necessary function to our llama farm, all feed and other costs associated with the cats are tax deductions.

As long as you adequately feed a cat, they will do fine without any special consideration when it is cold assuming you have a hay barn or barn where grass or hay is stored. However, since it is dry in barns, cats have the annoying habit of burying their feces inside the barn during the winter. This causes an odor.

A word of caution is necessary about cats, especially kittens. Cats can become infected by toxoplasmosis by eating infected small mammals. After becoming infected, they have infective toxoplasma in their feces from about 10 days after becoming infected to about two weeks. The cat builds up an immunity at that time and can no longer infect other animals. However, the parasite may live in the feces for up to a year. If your pregnant female llama comes in contact with the feces by eating contaminated feed, your llama may abort. Usually animals are immune after becoming infected one time.

Females of all types, including humans, can become infected by toxoplasmosis with similar results, that is abortion. Cats are most likely to become infected when they are young as older ones quickly build up an immunity. Therefore, you may want to consider having your females neutered so that there will not be a continual progression of kittens on your farm.

Cats may help your llamas adapt to having smaller animals around. This might help them to adapt to a new home with other types of animals. Cats are interesting critters and can be valuable assets for your llama operation. We and our llamas certainly enjoy our cat.

CHAPTER 33

SELECTIVE PUBLICATIONS AND INFORMATION ON LLAMAS

Major Llama Organizations:

International Llama Association, P O Box 37505, Denver, Colorado 80237. $50 per year for individuals and $75 for farm membership.

Llama Association of North America, P O Box 1882, Minden, Nevada 89423. $25 per year for individuals and $55 for breeder membership.

Rocky Mountain Lama Association, 15251 6100 Road, Montrose, Colorado 81401. $30 per year for individuals and $50 for ranch membership.

International Llama Association affiliate chapters:

Alaska Chapter, HC 31, Box 5247-A, Wasilla, Alaska 99517.
California Chapter, 11807 Simmerhorn Road, Galt, California 95632.
Central States Chapter, 6873 Haskley, Manchester, Michigan 48158-9711.
Greater Appalachian Llama Assn., RD 2, Box 738, Annville, Pennsylvania 17003.

Llama Assn. of Mid-Atlantic States, Rt. 2, Box 364F, Orange, Virginia 22960.

Llama Assn. of Southern California, 10711 Meads, Orange Park Acres, California 92669.

Llama Association of Washington State, P O Box 8585, Port Orchard, Washington 98563.

Northern Rockies Chapter, 7626 North 5th West, Idaho Falls, Idaho 83402.

Ohio River Valley Llama Assn., 5505 Dutch Lane, Johnstown, Ohio 43031.

South Central Llama Association, Route 1 Box 335, Fredericksburg, Texas 78624.

Sunshine States Llama Assn., Rt. 1, Box 279, Willianston, South Carolina 29697.

Willamette Valley Llama Assn., 4256 Roberts Ridge Road S., Salem, Oregon 97302.

Other llama organizations:

All Llamas of Hawaii Assn., P O Box A. Y., Kailua, Kona, Hawaii 96745.

Central Oregon Llama Assn., P O Box 5334, Bend, Oregon 97708.

Idaho Llama Breeders Network, 7626 North 5th West, Idaho Falls, Idaho 83702.

Iowa Llama Assn., RR 1, Box 175, Sioux City, Iowa 51108.

Klamath Basin Llama Assn., P O Box 7894, Klamath Falls, Oregon 97602.

Llamas of Minnesota, RR2, Box 68, Chatfield, Minnesota 55923.

Missouri Llama Assn., P O Box 467, Chillcouth, Missouri 64601.

New Mexico Llama Lovers, P O Box 3474, Pojoaque, New Mexico 87501.

Oklahoma Llama Assn., 126 N. Circle, Weatherford, Oklahoma 73096.

Wisconsin Organization of Llama Enthusiasts, 11308 Sherman Road, Cedarburg, Wisconsin 53012.

Magazines:

Canadian Llama News, 6012 Third Street S.W., Calgary, Alberta T2H 0H9. Bimonthly publication. $15 per year Canadian or U.S. funds.

Llama Life, P O Box 120, Mancos, Colorado 81328. Published quarterly, $16 per year.

Llamas Magazine, P O Box 100, Herald, California 95638. Published eight times per year, $25 per year, Canada $30 USA funds.

The Backcountry Llama, 2857 Rose Valley Loop, Kelso, Washington 98626. Published six times per year, $6.00 per year.

Llama Registration:

International Lama Registry, P O Box 7166, Rochester, Minnesota 55903.

Llama Show Association:

American Llama Show Association, Inc., P O Box 107, Ocate, New Mexico 87734.

PUBLICATIONS AND INFORMATION 262

Books:

Daugherty, Stanlynn. 1989. Packing With Llamas. Juniper Ridge Press, P O Box 338, Ashland, Oregon 97520. $13.95 postpaid.

Fowler, Murray. 1989. Medicine and Surgery of South america Camelids: Llama, Alpaca, Vicuna. Iowa State University, 2121 S. State Avenue, Ames, Iowa 50010. $76.95 postpaid.

Freeman, Myra. Heat Stress: Prevention, Management and Treatment in Llamas. Sunshine States Llama Association Marketplace, Rt. 1 Box 346, Lula, Georgia 30554. $15 postpaid (I haven't read this one).

Hoffman, Clare and Ingrid Asmus. 1989. Caring For Llamas: A Health and Management Guide. Rocky Mountain Lama Association, 168 Emerald Mountain Court, Livermore, Colorado 80536. $20.40 postpaid.

Markham, Doyle. 1990. Llamas Are The Ultimate; Training, Feeding, Packing, Hunting, Fishing and Care. Snake River Llamas, 7626 North 5th West, Idaho Falls, Idaho 83402. $16.95 postpaid. Idaho residents add $0.75 tax. Quantity discounts available.

Schreiber, John. 1988. Today And Tomorrow Llama Investment Analysis. Schrieber's Stoney Brook Farm, S9025 Highway E., Mukwonago, WI 53149. $14.95 postpaid.

Tillman, Andy. 1981. Speechless Brothers. Out of press, but copies may be available at a library or friend.

LLAMAS ARE THE ULTIMATE

The above books on llamas are ones that I can recommend. There are several others. Some I haven't read so I can't recommend them. Others that I have read aren't particularly useful to most llama folks. For a list of other books, you may send $0.50 to the International Llama Association for a brochure entitled, "Books, Videotapes, and Magazines About Llamas & Alpacas."

Llama Equipment And Products:

Electric Fence Systems:

Dare Products, Inc., P O box 157, Battle Creek, MI 49016.
Gallagher Power Fence. Inc. P O BOX 708900, San Antonio, Texas 78270.
Kiwi Fence Systems, Inc., Route 2 Box 51A, Waynesburg, Pennsylvania.
Live Wire Products, 1127 E Street, Marysville, California 95901.
New Zealand Fencing Systems, P O Box 518, Boring, Oregon 97009.
Northwest Fence Co., Rt. 1 Box 35, Enterprise, Oregon 97828.
Premier Fence Systems, P O Box 89, Washington, Iowa 52353.
Techfence, P O Box A-PN, Marlboro, New Jersey 07746.
Waterford Corporation, 404 N. Link Lane, Ft. Collins, CO 80524.
West Virginia Fence Corp., U S Route 219, Lindside, West Virginia 24951.

PUBLICATIONS AND INFORMATION 264

LLAMA PRODUCTS:

The following list of companies offering llama products were obtained from the national llama organizations. A questionnaire was sent to each company asking them to indicate the type of llama products offered. The companies with numbers in front of their names responded to the questionnaire. The numbers match the product category below:

1) Llama saddles, panniers and tack
2) Books or videos on llamas
3) Brushes, combs, etc.
4) Scales, Restrainers
5) Decorations, memo pads, forms, mugs, gifts, sculptures, etc.
6) Shirts, coats, ties, etc. with llamas and llama messages
7) Clothes made from llama wool, dyes, spinning equip., etc.

COMPANIES

2:Barkman Animal Enterprises, 34190 Lodge Road, Tollhouse, CA 93667, 209-855-6227

1:Backcountry Llamas, P O Box 1287, Paonia, CO 81428, 303-527-3844

1:Bonny Doom Llamas, 121 McGivern Place, Santra Cruz, CA 95060, 408-426-8649

Canyon Ranch, 17575 Jordan Road, Sisters, OR 97759, 503-548-6544

Coffman Farms, RR 1 Box 42, Roann, IN 46974, 317-833-5582

1,5:Columbia Crest Llamas, 415 Confer Rd., Kalama, WA 98625, 206-673-4477

5:Donna's Llama Stuff, P O Box 1529, Quncy, CA 95971, 916-836-2887

LLAMAS ARE THE ULTIMATE 265

5:Dragon Haven Llama Ranch, 980 Parkhill Road, Santa Margarita, CA 93453, 619-468-3269

2:Faik's Alpacas and Llamas, P O Box 521152, Big Lake, AK 99652, 907-892-8482

Foothill Llama & Alpaca Ranch, 460 Foothill Road, Hallister, CA 95023, 408-637-8655

Gilmore Llamas, P O Box 26, Murphy, OR 97533, 503-862-2407

1:Golden State Outfitters, 38475 Tack Road, Temecula, CA 92390, 714-676-6163

Hanta Yo Llamas, 5967 12th Ave., Sacramento, CA 95820, 916-457-9618

Hearthside Llamas, 11069 E. Wrenwood Ln., Clovis, CA 93612, 209-299-8057

Jane Marek, E. 4306 Sorrel, Spokane, WA 99207, 509-466-0153

2:Juniper Ridge Press, P O box 338, Ashland, OR 97520, 800-869-7342

Kesling's Llamas & Alpacas, 3300 Tally ho Drive, Kokomo, IN 46902, 317-453-7070

5,6:L. A. Llamas, 10825 Farralone Avenue, Chatsworth, CA 91311, 818-709-0575

Lla-Ess's, 9900 Wheatland Ave., Sunland, CA 91040, 818-352-3078

Llama Villa, P O Box 425, Garden Valley, CA 95633, 916-333-4866

1,5:Llamas of Michigan, 3540 76th St., Calidonis, MI 49316

Llamas of San Diego, 19526 Rancho Ballena Rd., Ramona, CA 92065, 619-789-7944

2,7:Llamas Las Medinas, P O Box 292, Sunal, CA 94586, 415-862-2028

1-7:Llamas & More, 6615 SW McVey, Redmond, OR 97756, 503-548-5821

5:Llsa Llama Ranch, 11383 Yuka Ridge Dr., Nevada City, CA 95959, 916-272-2834

1:Mama's Llamas, P O Box 655, El Dorado, CA 95623, 916-622-2566

Mary's Llamas, 3126 Pauline Drive, Sebastopoe, CA 95472, 707-823-8081

2:Meadowlark Farm Alpacas, 12305 NW McNamee Road, Portland, OR 97231, 503-283-2524

1,2,3:Mt. Sopris Llama Packs, 0270 Country Road 111, Carbondale, CO 81623, 303-963-3604

Pet-A-Llama, 5505 Lone Pine Road, Sebastopoe, CA 95472, 707-823-9395

1,3,4:R Llamas, P O Box 6619, Boise, Idaho 83707, 208-336-8842

Rancho Utopia, 5257 Maple Road, Vacaville, CA 95687, 707-448-4761

7:Real Cool Ranch, 29785 SW Burkhalter Rd., Hillsboro, OR 97123, 503-681-8333

1:Rocky Mountain Llamas, 7202 N. 45th St., Longmont, CO 80503, 303-530-5575

Sharon Mack, Rt. 1 Box 192, Elk, WA 99009, 509-292-8606

Sharp Llamas, 1855 Tomales Road, Petaluma, CA 94952, 707-778-1153

1:Shasta Packs, P O box 1137, Mt. Shasta, CA 96067, 916-926-3959

6:Shirt Stop, P O Box 639, Sisters, OR 97759, 503-549-9261

2:Silesian Llama Ranch, 2500, Box 1270, Frazier Park, CA 93225, 213-829-1287

6:Silver Mountain Prints, 35667 SE Squaw Mtn Road, Estacada OR 97023, 503-630-4033

1,2:Snake River Llamas, 7626 North 5th West, Idaho Falls, Idaho 83401, 208-524-0330

Snowmass Llamas & Alpacas, 0646 Capital Creek, Snowmass, CO 81654, 303-927-4317

2,7:Switzer-Land Alpacas & Llamas, Box 3800, Estes Park, CO 80517, 303-586-4624

The Conquering Llama, 4803 E-Soda Rack, Healasburg, CA 95448, 707-431-8457
6:The Spinning Llama, 10028 South Turkey Creek, Morrison, CO 80465, 303-697-8063
1:Timberline Llamas Inc., 30361 Rainbow Hills Road, Golden, CO 80401, 303-526-3876
1:Tiler's Llama Packs, 3266 Foothills Drive, Ronan, MT 59864, 406-675-4400
Two Bridges Llamas, RR 2 Box 110, Redfield, SD 57469, 605-472-0316
Volcano View Ranch, 18409 NE 10th Avenue, Ridgefield, WA 98642, 206-887-3636
6:Willow Creek Ranch Llamas, Star Route, Monument, OR 97864, 503-934-2963
Windmill Llamas, 1112 Jennings Road, Spangle, WA 99031, 509-624-0521
7:Windy Acre Llama Ranch, 9546 Glory Drive SE, Olymipia, WA 98503, 206-491-2177
1,3:Wildwood Llamas, Wildwood Acres, Allenspark, CO 80510, 303-747-2702
Yama Farms, 15651 SE Misty Drive, Clackamas, OR 97015, 503-658-5338

AT THE TRAIL HEAD

I hope that this book has increased your appreciation of llamas. If in some way, it has entertained, educated or benefitted your llama operation or prompted you to increase your contact with llamas, then I consider it a success. If llamas bring to your life a third of the satisfaction, enrichment and joy that our llamas have provided for us, then you're in for the time of your life! See you at the trail head....

LLAMAS REALLY ARE THE ULTIMATE!

ACKNOWLEDGEMENTS

Versions of portions of chapters 18, 24, 25, 28 and 29 have occurred in one or more of the following publications: Llamas Magazine, Llama Life, Idaho Outdoor Digest, Fins & Feathers, Sheep Magazine and Wyoming Wildlife.

I appreciate the assistance of Cherrie Markham, DeWayne Cecil and Larry Hailey for proof reading the final draft of this book.

All of the photographs appearing in this book were the work of either Doyle or Cherrie Markham with the following exceptions. Appreciation is extended to the following individuals for use of pictures: William Franklin, page 10; Steve Rolfing, page 12; Ralph L. Uber, page 13. The illustration on page 9 is by John Dawson, courtesy of the Natural History Museum of Los Angeles county.

INDEX

4-H 6, 193
adult iii, 5, 25, 27, 32-34, 36, 37, 52, 61, 71, 91, 99, 139, 140, 194, 195, 244, 246
africa 8
age iii, 16, 18, 45, 50, 51, 53, 56, 58, 89, 90, 106, 207, 239, 246
airplane 177, 195, 219, 220, 219
alfalfa 133, 138, 143, 144
alfalfa pellets 143, 144
alpacas 11-14, 262, 263, 265, 266
animals 2, 4, 8, 9, 11, 12, 27, 48, 51, 53, 54, 56, 59, 60, 70, 72, 77, 79, 82, 83, 84, 86, 88-92, 97, 103, 104, 125, 126, 129, 130, 132, 134-137, 139, 142, 145, 151, 154, 157-159, 162-165, 167, 170, 175-177, 187, 208, 217, 229, 230, 234, 238, 239, 244, 246, 248, 251, 258
antibiotics 249
antlers 53, 110, 171, 173, 193
archery 181, 190
Argentina 11, 13
arid 161
asia 8
axe 118, 123
babies 15-18, 32, 38, 87, 133, 135, 139, 140, 145, 146, 183, 191, 195, 216, 234, 244, 257
back iv, 14, 21, 26, 28, 34, 37, 38, 41, 42, 46, 48, 61, 62, 70, 72, 79, 83, 86, 88, 98, 99, 101, 111, 115, 116, 119, 125, 130, 141, 142, 171-173, 175-178, 181, 188, 191, 193-195, 197, 198, 201, 214, 216, 218-220, 222-224, 226, 228, 239-242, 248-251, 253, 254, 257
back packing 116, 197
backing 98, 134
barbed wire 156, 157
barley 144
barn iv, 7, 19, 137, 150, 151, 152, 154, 170, 255, 258, iv
barrier 42
bear 126-129, 128, 172, 177, 187, 212
behavior 1, 15, 29, 38, 48, 72, 108, 111, 125, 127, 170, 190, 192, 209, 246
bermuda grass 163
berserk male syndrome 15
Betadine 249, 254
big v, 6, 22, 26, 51, 73, 74, 84, 138, 143, 171, 176, 177, 181, 187, 191, 197, 201, 220, 221, 223, 226, 228, 249, 265

Big Horn Craigs 177, 191, 223
binoculars 124, 130, 225
bite 36, 133, 141, 182, 183, 192, 252
blower 60
boat 90, 195, 219, 227, 226, 228
bobcat 159
bog 38
boggy 65, 66
Bolivia 11, 13
books 154, 262-264
boulders 39, 66, 67, 179, 180, 226
breast bone 141
breast strap 77, 112, 111, 112, 178
britchin strap 77, 112, 114
browsers 139
bruises 250
brush 49, 60, 123, 167, 241
bucket 69, 116-118, 123, 133, 136
bumper hitch 95, 96
Bureau of Land Management 102, 167
buying 88
calm 18, 41, 69, 129, 132, 178, 190, 197, 223, 226, 228, 248
camel 8, 9, 125
camelidae 8
camp iii, 43, 75, 79, 80, 82, 103, 117, 121, 126, 128-131, 142-144, 171, 172, 173, 185, 189, 193, 199, 201
camper top 6, 95, 100, 101
camping 75, 80, 82, 107, 108, 116, 117, 123, 126, 131, 176, 192, 193
Canada 14, 150, 198, 212, 229, 261
canoe 177, 195
care i, iii, iv, vi, 4, 15, 106, 142, 149, 203, 217, 229, 232, 238, 262, 272
cat iv, 215, 255-258
catch 19, 25, 32-34, 47, 46, 50, 91, 125, 130, 131, 133, 135-137, 175, 234
catch pen 18, 19, 33
cattle 128, 143, 150, 238 (see cows)
Chamberlain Basin 177, 220, 224
Cherrie v, vi, 22, 24, 43, 46, 51, 54, 63, 67, 86, 112, 128, 140, 176, 177, 183, 185, 187, 192, 194, 197, 199, 202, 243, 244, 245, 257, 269, 271, 273
chest 117-119, 123, 142, 175, 199, 240, 252, 255

INDEX

children vi, 6, 193, 233
Chile 11, 13
chlorine 121
cinches 47, 46, 111, 178, 242, 251
clearance 98
clients 71, 89, 102-108, 157, 229, 231, 232, 234
climate 9, 13, 60, 147, 148, 150-155, 163, 179, 239, 251
clothes 6, 11, 12, 106, 115, 119, 179, 193, 264, 11
cobalt 148
cold 11, 39, 96, 115-118, 136, 140, 148-154, 175, 177, 194, 199, 201, 224, 233, 258
commercial packing iii, 102, 104
companies 57, 98, 158, 264
condition 41, 42, 53, 55, 58, 59, 62, 140, 141, 145, 164, 234
conformation 86
conifers 152, 175
cooking 107, 116-119, 123, 194, 199
copper 148
cord 43, 45, 91, 110, 131, 176, 193
corn 144, 176, 194, 201
corral 18, 19, 21, 29, 33-35, 45, 71, 133, 135, 136, 137, 139, 140, 144, 145, 147, 151, 152, 158, 159, 160, 162, 165, 166, 168, 170, 182, 212, 213, 215, 246, 257
cost 5, 38, 57, 90, 92, 96, 99, 101, 105, 107, 108, 145, 207, 234, 235, 257
cow-hocked 88
cows 88, 146, 168, 195 (see cattle)
coyote 159, 206, 207, 209-212, 214, 215, 216
creek 38, 39, 42, 43, 75, 131, 187, 192, 215, 218, 225, 226, 263, 266, 267
cria v, 16 (see babies)
croplands 161
crossbars 46
culvert 42, 43
curious 127, 128, 178, 195, 252, 257
cut 138, 147, 157, 187, 212, 213, 224, 230, 239, 242-245, 247, 248, 249, 250, 254
cutthroat 201
deciduous trees 152
deer 49, 54, 84, 128, 172, 175-177, 181, 185, 187, 188, 187, 190, 191
defecates 80
defecating 77
dental 239
desensitized 45

disease 4, 14, 33, 179, 206, 238
dog 5, 42, 79, 80, 127, 145, 159, 192, 206-208, 211, 212, 214-216
domestic stock 11, 13, 140, 150, 156, 165, 176, 183
drink 57, 68, 69, 75, 115, 121, 122, 143
drought 169
dung 153, 165, 169
dust 60, 170
ears 22, 88, 178, 216, 218, 239, 253
east 66, 154, 189
eat 11, 43, 63, 138-140, 144, 146, 152, 162, 165, 167, 178, 195, 217, 257
economical 4, 6, 148, 161, 195, 201, 205, 208
electric 99, 158, 159, 158-160, 165-170, 212, 263, 158
elk 4, 49, 53, 54, 71, 75, 84, 110, 128, 172, 173, 175, 176, 179, 187, 189, 190-192, 194, 224, 225, 266
emergency iv, 98, 130, 232-235, 248
energy 138, 146
environment 80, 81, 82, 149, 150, 176, 190, 195, 206
environmental disturbance 190
etiquette iii, 79
eyes v, 1, 61, 66, 99, 183, 193, 197, 214, 228, 252, 253, 254
eyesight 4
fall 53, 68, 72, 96, 115, 116, 142, 143, 147, 152, 161, 167, 177, 178, 192, 193, 196, 201, 218, 219, 224, 239-241, 257
farm vi, 15, 26, 33, 39, 43, 70, 71, 89, 101, 138, 146, 148, 159, 161, 170, 175, 192, 193, 195, 206, 235, 256, 255, 257, 258, 259, 262, 266
fat 141, 142, 149
feces 6, 147, 164, 165, 258
feed 5, 7, 33, 63, 75, 121, 123, 130, 133, 138-141, 143, 142, 144-149, 161, 179, 182, 195, 217, 218, 255, 257, 258
feeder 17, 146, 147
feeding i, iii, iv, 15, 17, 57, 138, 139, 141, 143-147, 149, 162, 183, 225, 232, 255, 262, 272
feet 11, 13, 17, 20, 22, 24, 28, 38, 41, 55, 88, 96, 126, 127, 148, 158, 165, 166, 169, 176, 178, 179, 183, 193, 212, 214, 223, 224, 242, 253 (see foot)
females 5, 6, 15, 17, 21, 22, 24, 27, 32, 34, 38, 70, 71, 97, 101, 127, 133-135, 139, 140, 142, 145, 146, 149, 154, 156, 158, 160, 167, 181-183, 191, 194, 195, 210, 216, 233, 244, 246, 253, 255, 258
fence iv, 6, 17, 34, 127, 129, 133, 134, 139, 156-161, 166-170, 179, 192, 195, 208, 212, 213, 215, 243-246, 263

fertilizer 6, 145, 146, 169, 170, 182
fifth-wheel hitch 95
fight 36, 156, 158, 159, 208, 239, 240, 242-247
fire 80, 103, 117-119, 121, 123, 199, 201, 225, 226, 80
fire pan 103
first aid 107, 118, 123, 248, 254
first aid kit 118, 254
fish 103, 119, 183, 197, 200, 201, 204, 205
fishing i, iii, iv, 102, 124, 183, 197-199, 202, 201, 203-205, 248, 254, 262, 272
flashlights 121, 123, 129, 192, 121, 129
flies 124, 252
flying iv, 197, 199, 218, 219, 221, 222, 228
food 5, 12, 107, 115, 117-119, 139, 155, 176, 187, 194, 199, 204, 205, 217, 255, 257
foot 25, 29, 38, 42, 44, 43, 88, 95, 96, 98, 99, 106, 110, 114, 127, 130, 131, 136, 144, 148, 156, 165, 167, 169, 176, 185, 190, 191, 193, 197, 219, 224, 226, 228, 242, 249, 252 (see feet)
forage 63, 75, 138, 140, 142, 143, 162-166, 170
Forest Service 102, 105, 176, 190
four-wheel drive 6, 26, 95, 100, 171, 173
Fowler, Murray 142, 238
fox 159
fracture 250
fragile 80, 81, 150, 176, 190
Frank Church River of No Return Wilderness 68, 177, 223
frost 145, 148, 201, 224
fur 1, 179
furacin 254
game iv, 1, 4, 49, 91, 103, 110, 115, 171, 173, 175, 176, 183-185, 187
geldings 7, 15, 23, 25, 26, 29, 45, 70, 71, 135, 137, 170, 175, 211, 219, 245-247, 250
goat 7, 92
gourmet 107, 117, 177, 194, 199
grain 75, 76, 121, 123, 130, 133, 136, 140-144, 148
grass 133, 138, 144, 146, 162, 163, 165-167, 169, 170, 178, 195, 218, 224, 258
grass hay 133, 138
gravel 136
grazers 76, 139
grazing 5, 45, 76, 81, 140-142, 144-146, 161-167, 169, 170, 176, 203, 224
green llama 49, 65, 71, 72

grizzly bears 82, 83, 128, 212
guanaco 8, 11, 12, 11, 12, 209, 210
guarding iv, 7, 151, 161, 208, 209, 211, 213, 272
guide 103, 105-107, 121, 131, 161, 262, 103, 108
guiding 108
halter 19-22, 25, 27, 32-36, 35, 45, 107, 123, 132, 134, 137, 145, 183
haltering 33, 135, 136
hands 17, 21, 33, 66, 115, 119, 134, 135, 179, 184, 189, 218, 223, 226
harem 210, 240
hay 5, 17, 24, 27, 57, 97, 133, 138-140, 145-148, 152, 154, 161, 162, 179, 182, 183, 195, 215, 217, 255, 258
health iv, 24, 92, 140, 141, 149, 166, 191, 217, 229, 232, 234, 238, 243, 247, 262
health certificate 24, 191, 234
heat 62, 66, 100, 148, 151, 154, 251, 253, 262
heat stress 62, 154, 251, 253, 262
height 26, 27, 56, 84, 86, 100, 163, 223, 243
herd 6, 18, 33, 34, 71, 76, 125, 126, 128, 133, 136, 137, 167, 183, 213, 229, 232, 236
hide 11, 127, 172, 187, 193, 218
history iii, 8, 84, 269
horses 8, 79, 81, 92, 94, 114, 127-129, 143, 150, 168, 176, 178-182, 189, 210, 218, 219, 233, 251
humidity 66, 151, 153, 154, 251
hunt 4, 71, 72, 183, 192
hunting i, iii, iv, 4, 8, 26, 53, 70, 75, 102, 110, 115, 124, 127, 142, 143, 171, 172, 176, 177, 182, 183, 185, 186, 185, 187, 189, 190-192, 196, 199, 248, 262, 272
hyperthermia 251
ice 8, 117-119, 123, 149, 176, 179, 194, 199, 218, 224, 250
ice chests 117, 176, 194, 199
Incas 11, 13
Indiana 232
injection 239, 244, 247, 253, 239
injuries 41, 104, 108, 248, 249, 253, 254
insects 203
intelligent 93, 177, 183, 192
intermountain west 99, 179, 251, 272
investment 6, 86, 208, 209, 217, 229, 262
iodine 121, 148, 254
iron 119, 148
irrigation 39, 145, 157, 161, 163, 168
ivermectin 239, 244, 252, 253, 255
javelina 49, 187

INDEX 276

Rais, Jay 92, 125
Jebidiah Smith Wilderness 192
jet boat 90, 219, 226, 228
jogging 182
Johnson, LaRue 142, 254
jump 6, 20, 24-27, 36, 38, 39, 41-43, 46, 48, 63, 66, 77, 91, 100, 101,
115, 129, 131, 156, 173, 175, 177, 191, 213, 219,
223, 225, 226, 228
juvenile 27, 135, 140
kick 27, 32, 33, 37, 80, 88, 176, 182, 183, 192, 27
knapweed 167
knees 29, 88, 220
knocked knees 88
kush iii, 29, 30, 29-31, 222, 227
lakes 80-82, 176, 196, 198, 197-200, 202, 201, 203, 204
lama 8, 10-12, 138, 259, 261
lantern 121, 123
lay 27, 28, 60, 185, 194, 213
lead 19-24, 27-29, 32-39, 41, 42, 46, 48, 50, 59-63, 66, 68, 70, 72-75,
77, 89-91, 101, 106, 107, 110, 111, 123, 126, 134,
136, 178, 183, 191, 198, 199, 216, 225, 228, 232,
244, 245, 250
learn 6, 21-26, 28, 36, 37, 38, 42, 43, 45, 49, 52, 62, 66, 98, 177, 191,
192, 195, 216, 221
leather 17, 19, 35, 37, 114, 185, 222, 250, 254
leg 21, 29, 30, 36, 37, 39, 88, 230, 249, 250, 253
legs 21-24, 26, 29, 30, 36, 37, 39, 45, 46, 61, 87, 86, 88, 111, 119, 141,
158, 178, 193, 220, 222, 226, 230, 239, 240, 244,
246, 248, 249-253
lie 6, 29, 30, 37, 45, 46, 56, 59-62, 70, 97, 99, 101, 151, 177, 181, 193,
194
lie down 6, 29, 30, 37, 45, 46, 56, 59-62, 70, 97, 99, 101, 151, 177,
181, 193-195
ligament 250
limb 42, 199
lion 125, 126
livestock 6, 56, 57, 60, 83, 99, 101, 103, 139, 140, 148, 151,
156, 163, 164, 165, 179, 206, 239, 252,
257
llama organizations 259, 260, 264
llama restraints 242
load v, 1, 6, 23-28, 32, 36, 49, 50, 53-56, 59, 60, 74, 84, 86, 88, 91, 96,
97, 112, 114, 141, 172, 175, 176, 178, 181, 186, 187,
191, 220, 223, 234, 249-252

loading 24-28, 36, 56, 58, 97, 98, 173, 187, 223
lock 97, 118, 119
lost iv, 7, 125, 130, 131, 207, 209, 211, 212, 216
lung 89
magazines 261, 263, 272
management iv, 49, 102, 145, 146, 161, 162, 164, 165, 167, 169, 170, 236, 262, 272
manager 29, 147
manganese 148
manger 97, 182
mature 5, 38, 50, 52, 55, 56, 70, 141, 163, 179, 181, 191, 233, 240, 246, 255
maximum 53, 54, 56, 96, 171, 186, 191
maximum weight 53, 56
meals 106, 116-119, 176, 177, 194, 199
medicine 229, 232, 236, 239, 250, 262
medium 60, 84, 86, 98, 101, 179
metabolize 140
mice 255, 257
minerals 148
mini pickup 99, 181, 194
minivans 101, 194
Montana 26, 116, 127, 128, 130, 172, 173, 177, 194, 199, 232, 253
moose 26, 49, 75, 84, 128, 171-173, 175, 179, 187, 191
mosquito 115, 252
mountain 6, 11, 38, 43, 45, 48, 49, 59, 66, 77, 78, 80, 86, 98, 100, 125, 126, 127, 130, 131, 141, 142, 171, 176, 181, 196, 197, 199, 200, 202, 203, 205, 221, 259, 262, 266
mouse 255, 257
mud 88, 126, 176, 241, 65
muddy 65, 189
muscles 22, 56, 88, 141, 145, 239
national forest 82, 83, 128, 131, 143, 176, 191, 231, 264
neck 19, 21, 28, 30, 35, 42, 45, 59, 63, 111, 131, 134, 214, 221, 222, 239, 240, 243-245
New Mexico 187, 232, 260, 261
New Zealand 158, 159, 158-160, 164, 166, 168, 169, 263
night v, 11, 13, 29, 45, 63, 66, 77, 96, 116, 121, 125, 126, 140, 143, 150, 176, 177, 194, 201, 206, 208, 209, 212, 215, 216, 226, 233
North America 8, 9, 8, 11, 13, 14, 70, 150, 161, 175, 191, 198, 199, 203, 206, 224, 259
nose 35, 244, 252
nursing 38, 140, 146

INDEX

nutrient 138, 163
nutrition 138
oats 144
odor 6, 170, 182, 258
old 15, 17, 22, 24, 25, 39, 43, 46, 50-52, 54, 56, 68, 88, 90, 125, 126, 131, 148, 156, 171, 172, 175, 180, 181, 183, 191, 192, 195, 208-210, 233, 239, 246, 255, 273
ophthalmic 254
outfitters 102, 103, 105, 106, 107, 118, 265
outfitting 105, 107-109
over-weight 84, 140
overgrazing 81, 164, 170
overload 50
pack iii, 1, 6, 38, 43, 45, 46, 48-52, 55, 56, 59-61, 63, 65, 67, 66, 70, 71, 72, 73, 75-82, 84, 86, 88-90, 92, 95, 97, 98, 102, 103, 105, 108, 110, 111, 114, 115, 117-119, 125, 126, 129, 130, 132, 139, 144, 172, 173, 175, 176, 178, 179, 181, 183, 185, 186, 188, 187, 190-193, 195, 197, 199, 203, 208, 213, 217-219, 246, 248, 251, 252
pack animals 48, 49, 79, 83, 84, 86, 88-90, 103, 130, 132, 139, 175, 176, 179, 187, 192, 193, 208, 217, 248
pack stock 70, 79, 81, 82, 183, 197, 219, 248
pack string iii, 43, 50-52, 63, 65, 70-73, 75, 77, 78, 89, 90, 125, 176, 178, 195, 246, 248
packers 1, 4, 14, 30, 50, 51, 60, 63, 72, 75, 77, 79, 83, 84, 86-90, 92, 112, 125, 128-131, 177, 178, 180, 181, 229, 248
packing i, iii, iv, 1, 4, 6, 13, 38, 49-51, 50-53, 56-58, 60, 66, 69-73, 75, 76, 77, 79, 83-87, 86, 88-93, 95, 97, 99, 102, 104, 105, 108, 110, 111, 114, 116, 118, 119, 123, 127, 132, 140-145, 149, 171-173, 175, 176, 182, 183, 185, 187, 190-192, 194, 195, 197, 200, 223, 234, 246, 248, 249-252, 262, 271, 272
packing llama 84, 85, 140, 223
pads 46, 81, 88, 116, 123, 190, 241, 242, 249, 250, 264
panic 34, 49, 66, 79, 125, 131, 193
pannier 49, 72, 110-112, 115-119, 121, 187, 199, 218
panniers iii, 6, 38, 45, 46, 48, 49, 54, 55, 59, 60, 72, 86, 91, 106, 107, 110-112, 114-119, 121, 130, 173, 176-178, 187, 188, 192, 193, 199, 218, 223, 264
parasite 239, 258
Park Service 83, 102, 176, 190
pastern 88

pasture iv, v, vi, 5, 15, 17-19, 71, 133, 134, 133, 136-140, 144-147, 152, 156, 157, 160-162, 161-166, 165-170, 182, 195, 211, 212-215, 244, 246, 253, 255, 257
pasture management iv, 145, 146, 161, 164, 165, 167, 169, 170
pay 89, 90, 103, 107, 108, 127, 128, 235, 236
pellets 80-82, 140, 143, 144, 166, 169, 170, 176, 182
personalities 18, 90, 106, 178, 195, 231
Peru 11-13
photography 102, 203
pickup 6, 26, 95, 100, 99, 101, 181, 194
plants 6, 62, 163, 164, 167, 170, 182
plastic pipes 137
pocket 112, 115, 130
poles 35, 123, 135, 157, 243, 35
portable fence 103, 165-170, 176, 194
pounds 1, 22, 51, 54-57, 74, 85, 86, 106, 110, 114-116, 179, 187-189, 191
predators 7, 125, 158, 159, 206, 207, 209-211, 214, 216, 217, 257
pregnant 142, 182, 233, 258
prices 88, 90, 94, 195
products ii, 263, 264
purchasing 5, 7, 32, 57, 90, 91, 93, 96, 98, 101, 102, 105, 114, 119, 193, 194, 200, 212
rabbit pellets 140, 144
rain 106, 123, 151
rainbows 201
rainfall 157, 161, 163
range lands 164
Rattlesnake 218, 252
regulations 79
relieve 28, 181, 183
religious symbols 14
rest 20, 33, 35, 43, 62, 72, 75, 78, 116, 121, 139, 144, 145, 162, 164, 171, 250, 251
restrainers 242, 243, 264
rivers 203
road 55, 98, 100, 101, 115, 118, 171, 176, 183-185, 189, 193, 194, 219, 228, 259, 260, 261, 264-267
roadless area 114, 194, 198, 197, 198, 203
rocks 41, 73, 115, 121, 193, 226
rope 19, 27, 28, 33, 35, 45, 63, 72, 73, 81, 107, 110, 114, 130, 134, 133, 134, 135, 137, 191
rotation 138, 146

INDEX   280

rotational grazing 164-167
rule 15, 79, 83, 84, 105, 106, 139
ruminants 139
run 19, 21, 50, 60, 108, 119, 121, 125-127, 131-133, 182, 183, 211,
        214, 215, 219, 246, 257
saddle iii, 6, 21, 26, 34, 38, 45-49, 54, 59, 60, 63, 72, 74, 77, 78, 86,
        91, 107, 110-112, 114, 115, 117-119, 175, 177, 178,
        192, 223, 250, 251, 264
safe 2, 6, 34, 86, 121, 144, 150, 183, 192, 221, 243, 244
Sage Creek 39
Salmon River 90, 177, 219, 221, 224, 227, 226, 228
salt 148, 273
saw buck 72
scales 56, 57, 91, 92, 114, 141, 217, 243, 264
Schuh, Dwight 191
screw-in stakes 45, 126, 131
seed 88, 146, 161, 163, 167
selenium 148
semi-arid 161
session 20-22, 24, 37, 191, 221
shed iv, 150-154, 209, 215, 255
sheep iv, 6, 7, 11, 13, 60, 116, 146, 150, 151, 161, 165-168,
        170, 177, 206, 207-210, 209-217, 269, 272
shelter 111,112, 150-155, 179, 195
shooting 172, 178, 219
short duration grazing 164
shots 15, 18, 178, 243-245, 247, 252, 255
shovel 118, 123
shrubs 45, 139, 152
sleep 97, 116, 224
sleeping bag 50, 96, 106, 116, 119, 123, 128, 201, 233
slow 61, 62, 75, 78, 225
small 5, 9, 11, 12, 18, 19, 27, 33, 43, 63, 76, 77, 80, 84, 86, 91, 116,
        118, 119, 121, 130, 131, 133, 139, 141, 145, 153,
        154, 167, 168, 175, 177, 185, 187, 192, 199, 206,
        211, 212, 217, 222, 228, 230, 239, 245, 247, 255,
        257, 258
smart 4, 22, 37, 45, 91, 209
smell 6, 81, 89, 111
snow 8, 26, 65, 68, 117, 127, 133, 140, 142-144, 147, 151, 153, 171,
        172, 173, 175, 185, 189, 197, 199, 201, 203, 208,
        219, 221, 223, 224, 240, 241
solar shower 120, 121, 123

south 6, 8, 9, 11, 12, 11-14, 66, 84, 125, 138, 150, 153, 183, 187, 209, 260, 262, 267
South America 6, 8, 9, 11-14, 84, 125, 138, 150, 153, 183, 187, 262
spaniards 11
spit 17, 70, 71, 135, 182
splinting 250
spook 178, 181, 193
spring 60, 68, 130, 133, 141, 146, 162, 163, 167, 169, 203, 205, 213, 218, 219, 238, 240, 241, 252, 253
stake out 176
steep 55, 65, 89, 112, 172, 178, 180, 182, 218, 219, 225, 226, 228
stock rack 6, 26, 95, 99, 101, 103, 173, 175, 181
stock trailers 94, 95, 98
stomach 30, 139, 141, 251
storage 97, 101, 118, 154
stove 116, 123, 194
stream 38, 41, 42, 63, 64, 69, 77, 80, 82, 167, 170, 192, 199, 203, 204
stress 62, 151, 154, 229, 251, 253, 262
string iii, 42, 43, 50-52, 63, 65, 70-73, 72, 73, 75, 77, 78, 89, 90, 110, 125, 175-178, 195, 246, 248
studs 7, 14, 26, 38, 45, 70, 71, 73, 82, 97, 110, 135, 137, 139, 156-159, 175, 177, 178, 181, 182, 194, 210, 216, 219, 220, 240, 245-247, 249
summer 1, 4, 11, 26, 56, 62, 68, 80, 96, 100, 117, 128, 130, 133, 138, 139, 143, 144-146, 148, 150, 153, 163, 169, 175, 176, 183, 190, 192-194, 196, 199-201, 203, 215, 225, 241, 250, 257
suture 250, 254
teeth 239, 240, 242-245, 247 (see teeth)
temperatures 13, 140, 149, 151, 179, 185, 195, 203, 217, 251
tent 106, 110, 111, 116, 123, 224, 226
tetanus 239, 250
tether 44, 43, 45, 110, 123, 126, 128, 130-132
tether line 126, 131
thermometer 251
thighs 22, 88
thistle 167
ticks 252, 253
tie 35, 45, 46, 49, 50, 59, 60, 65, 71, 72, 75, 77, 82, 97, 101, 111, 114, 125, 135, 193
tied 19, 26, 27, 38, 42, 45, 54, 59, 61, 63, 65, 70-72, 75, 76, 81, 82, 101, 110, 112, 114, 119, 128, 130, 131, 154, 175, 178, 187, 191, 193, 199, 222, 224, 226, 233, 251

toe nails 92, 240, 241, 243-245, 255
tooth 8, 244, 245 (see teeth)
topographic map 177
toxoplasmosis 258
trace minerals 148
trail iii, iv, v, 28, 38, 41, 42, 49, 50, 59, 62-64, 69-73, 78, 79, 81, 80, 81, 82, 86, 91, 103, 107, 115, 121, 125-130, 132, 138, 141, 144, 178, 181, 182, 189, 192, 193, 199, 200, 204, 218, 219, 225, 226, 248, 250, 253
trail head iv, 59, 62, 70, 86, 107, 126, 130, 179, 199, 204, 250, 253, 268
trailer 6, 24-28, 36, 59, 70, 71, 94-101, 107, 173, 183, 191, 221, 234, 244, 251
train 5, 6, 18, 32, 34, 42, 133, 182, 191, 217
training i, iii, vi, 6, 15-20, 22, 24, 29, 31-33, 36-39, 42, 43, 45, 49, 59, 60, 61, 64, 63, 65, 72, 90, 91, 103-105, 134, 135, 139, 182, 183, 191, 209, 219, 221, 230, 232, 243, 244, 262, 273
tranquil 1, 18, 99, 172, 175, 208, 217, 221, 248
transport iii, 14, 24, 27, 59, 94, 95, 98-101, 103, 107, 181, 191
trees 38, 42, 43, 45, 70, 73, 75, 80, 81, 103, 110, 115, 127, 132, 139, 151, 152, 171, 175, 185, 187, 193, 221, 224-226
trek 26, 49-51, 55, 59, 73, 90, 103-108, 111, 115-118, 122, 121, 175, 181, 188, 199-201, 203, 205, 228, 250
trekking 45, 55, 59, 62, 71, 77, 78, 92, 103, 107-109, 115, 121, 123, 199, 203, 252
trout 197, 201, 202, 201, 204
truck 7, 23-27, 36, 54, 57, 59, 60, 76, 94-96, 99-101, 107, 133, 172, 173, 175, 177, 192, 214, 234
trunks 38, 42, 43, 115, 226
twine 46, 72, 131
undergrazing 164, 170
unload 49, 75
untrained 32, 33, 35, 96, 101, 135, 178, 183
urine 164, 165
useful life 195
vaccinating 238, 239
vaccine 238, 239, 247
van 6, 59, 94, 101, 181, 194
vehicle 91, 99, 181, 194, 197
vents 96, 100
veterinarian iv, 24, 147, 148, 229-236, 238, 239, 246-251, 253, 254, 257

veterinary 229, 233
vicuna 8, 10-12, 262
water 8, 38-40, 39, 41, 63, 68, 69, 75, 80-82, 91, 103, 110, 114, 115, 118, 122, 121, 123, 143, 144, 147-149, 166, 169, 176, 185, 197, 201, 203, 215, 225, 226, 228, 238, 245, 249, 251, 252, 254
water purification 121, 252
water troughs 148, 169
waterer 148
wean 18
weanling iii, 18, 21, 29, 32, 33, 37, 38, 135
weather 50, 96, 106, 116, 121, 149, 151-154, 169, 179, 206, 217, 251
weigh 1, 56, 57, 86, 99, 106, 114, 141, 191
weight 1, 37, 50, 53-58, 84, 86, 91, 92, 96, 98, 99, 106, 110, 115, 116, 119, 130, 139-142, 158, 179, 181, 191, 194, 207, 209, 249
wilderness 1, 4, 69, 76, 84, 120, 125, 126, 130, 132, 177, 191, 192, 194, 198, 197, 198, 200, 203, 219-221, 223, 224, 228
wildlife 128, 164, 201, 203, 205, 207, 269, 271
wind 73, 96, 99, 151-154, 179, 180, 194, 195, 200, 214, 217
Wind River Range 73, 179, 180, 194, 200
winter 7, 11, 96, 116, 133, 140, 142, 143, 146-148, 150-154, 157, 161, 167, 203, 217, 250, 258
women 179, 193
wood 59, 80, 111, 116, 243
wool 6, 11-13, 47, 46, 60, 84, 88, 91, 92, 135, 139-141, 147, 154, 157, 170, 179, 193, 195, 207, 217, 251, 264
worm 239, 240
wounds 248, 252, 254
Wyoming 70, 73, 128, 131, 179, 185, 189, 191, 192, 194, 199, 200, 269
Yellowstone National Park 131
youngster 18, 41, 50, 51, 55, 66, 191
zinc 148

NOTES

NOTES

MEET THE AUTHOR

Doyle Markham is a Research Ecologist for a federal agency and a Certified Wildlife Biologist. He and his wife, Cherrie, began raising, breeding and packing with llamas in 1983. Cherrie is a math teacher for a local school district. They live in Idaho Falls, Idaho and own and operate Snake River Llamas, one of the largest llama herds in the intermountain west. They do all the feeding, maintenance, care and marketing of their llamas. In addition to selling llamas, Snake River Llamas sells llama packing equipment.

As a free-lance writer and photographer, he has published many articles and photographs in outdoor magazines on a variety of subjects, including three magazine cover photographs. Llamas are one of his favorite subjects, and he has published extensively in llama-related magazines. He teaches llama packing courses for community education systems.

Dr. Markham has given numerous slide presentations on hunting, packing and fishing with llamas, sheep guarding with llamas and general llama management and breeding. He gave the after dinner speech and slide presentation at the New Friends-Old Friends Dinner at the 1989 International Llama Association Conference in Salt Lake City.

He and Cherrie have trained numerous llamas. They have packed, fished and hunted with llamas in many western states. For several years, Doyle and Cherrie have been co-editors of the Northern Rockies Chapter, International Llama Association, Newsletter, and he has written many training and other articles for the newsletter. In 1991, Markham will serve as the president of the Northern Rockies Chapter. He is also the president of the Idaho Llama Breeders Network.